The Origins of Inequality

Per Molander

The Origins of Inequality

Mechanisms, Models, Policy

 Springer

Per Molander
Affiliated to the Institute for Evaluation of Labour
Market and Education Policy
Uppsala, Sweden

ISBN 978-3-030-93191-9   ISBN 978-3-030-93189-6   (eBook)
https://doi.org/10.1007/978-3-030-93189-6

This Springer imprint is published by the registered company Springer Nature Switzerland AG.
The registered company address is: Gewerbestrasse 11, 6330 Cham, Switzerland

Preface

Increasing inequalities in the OECD countries and many other parts of the world have fuelled a debate on public policies in almost all areas relevant to the distribution of opportunities and resources in society. This debate has also sparked new or renewed interest in inequality in the social sciences. This book is an attempt to convey some of the basic insights from old and new research that could benefit a debate that tends to be governed by ideology and entrenched positions.

The scope of the undertaking sets strict practical limits to what can be covered. For most of the topics dealt with, there exist textbooks, handbook chapters, and survey papers that present the material in more detail. Such texts have formed the basis of the current work, supplemented in many areas with single scientific papers and reports. In this sense, the book can best be likened to an aerial photo of a vast terrain, with the aid of which the reader can identify areas that merit deeper interest and can be explored using the primary literature referred to.

It is inevitable that the focus of the presentation and the choice of examples is marked by the author's frame of reference. Most of the empirical evidence has been gathered in industrialised countries, but in some areas, the perspective is widened to the global arena. Within the OECD, there is some bias towards studies based on data from the USA and the Nordic countries.

This book has a fairly long history. I first began to think about these matters around 1980, when I had finished my PhD studies. KarlJohan Åström, my mentor at the time, suggested that I should learn economics. I started with a title that appeared appropriate for someone with prior knowledge on optimisation: *Mathematical Optimization and Economic Theory*, by Michael Intriligator. The book's sixth chapter deals with game theory, including John Nash's theory of bargaining. An exercise towards the end of this chapter in Intriligator's book is aimed at testing whether the student masters Nash's solution to the bargaining game and carries the message that a wealthy person will gain more from a bilateral negotiation than a poor person. The reflexes that I had acquired during my studies of control theory were aroused; here was an example of positive feedback, implying instability if the game of bargaining is repeated. I expected to find a chapter discussing how to control this instability, but there was none.

Three decades or so later, I found the time necessary to dig deeper into the problem of inequality, as well as the social science research devoted to it. There is a vast literature on both income inequality and social stratification—scientists such as Tony Atkinson, Amartya Sen, and Thomas Piketty have become known to a wider public—but the general focus of this work is on marginal change. Typical questions asked are why economic inequality is somewhat more pronounced in some liberal democracies than in others, or how the Gini coefficient of personal income distribution in a given country has developed over the last few decades, and so on. Relatively little energy has been spent on the roots of inequality. This book aims to fill part of this lacuna.

Uppsala, Sweden   Per Molander
September 2021

Acknowledgements

Part of the material presented was collected while I chaired the public commission on equality appointed by the Swedish government in 2018. For stimulating discussions and input to the report of the commission, I am indebted to its members (Anna Balkfors, Anders Björklund, Anna Hedborg, Helena Holmlund, Olle Lundberg, and Clas Olsson), experts (Mats Johansson and Ruth-Aïda Nahum), and the secretariat (Tove Eliasson, Marie Jakobsson Randers, Kristian Persson Kern, and Frida Widmalm). I am deeply grateful to Olle Lundberg, Martin Nybom, and anonymous reviewers, who all helped improve the quality of the text by commenting on earlier versions.

Over the years, I have learnt a lot from friends and colleagues in the social sciences. Among these, I would like to mention Anders Forslund, Torsten Persson, Bo Rothstein, and Jörgen Weibull.

As always, the author remains solely responsible for whatever flaws remain in the text.

About the Book

This book presents a unified approach to the emergence of inequality in human societies. The analysis is based consecutively on the individual life cycle, social mechanisms that generate inequality, and spatial inequality. For each of these perspectives, there is one chapter summarising relevant empirical research and a companion chapter on mathematical models used for analysis. As a result, the book is also accessible to readers with little or no mathematical training. A separate chapter is devoted to policy implications on a general level.

Contents

About the Author

Per Molander holds a Ph.D. in applied mathematics (control theory) but has spent most of his professional life at the interface between the social sciences and public policymaking. He has published around 100 articles, reports, and books, including *Turning Sweden Around* (MIT Press), *Institutions, Politics and Fiscal Policy* (Kluwer), *Fiscal Federalism in Unitary States* (Kluwer), *Alternatives for Welfare Policy* (Cambridge University Press) (all these co-authored), and *The Anatomy of Inequality* (Melville House/Random House). He was the main architect of the Swedish central government budget reform of the 1990s and created the Swedish Social Insurance Inspectorate, where he served as the director general between 2009 and 2015. He chaired the Public Commission on Equality that delivered its report in 2020. He has also worked as an adviser to the IMF, the World Bank, and the OECD. He has received several literary prizes, including the Essay Prize from the Swedish Academy. Dr. Molander is a member of the Royal Swedish Academy of Sciences.

Chapter 1
Introduction

Inequality is ubiquitous. Virtually all human societies are marked by inequality in the distribution of resources, at a level beyond that which could be expected due to differences in individual capabilities or effort alone. This is true not only for agricultural and industrial societies but also for social entities that are perceived as earlier stages in human social development,[1] and holds even if the perspective is extended to primate societies.[2]

Another general pattern that calls for an explanation is that inequality tends to grow over time. This development has been mapped for parts of Europe and the United States for preindustrial and modern history,[3] as well as, in less detail, in a global and long-time perspective.[4] An illustration is given in the figure below—the development of average inequality in Europe from the Middle Ages until the present day (Fig. 1.1).

The tendency for inequality to grow is not without exceptions. Wars and epidemics have reduced inequality during some periods, and trade unions and universal suffrage both contributed to a combination of sustained growth and decreasing inequality in large parts of the developed world between the late nineteenth century and the 1970s. But these periods appear to be exceptions to a general rule.

A third general observation is that individual achievements—which are often used in the public discourse to legitimate inequality—are socially determined. A well-paid professional such as the managing director of a company in an industrial society may be convinced that his or her salary simply reflects individual talent and effort, but if this person were moved to a developing country with a less well-functioning infrastructure, a generally lower level of education, pervasive corruption and other problems, his or her productivity would fall significantly. In fact, about

---

[1] Bowles et al. (2010).

[2] de Waal (2005).

[3] Alfani (2021), Lindert (2000), Morrison (2000).

[4] Bourguignon and Morrisson (2002), Scheidel (2017).

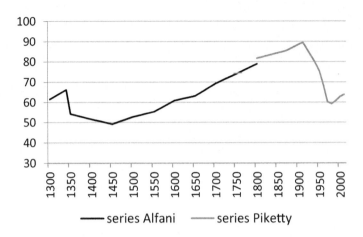

Fig. 1.1 Inequality in Europe 1300–2010. Share of wealth of the richest 10 per cent. Source: Data from Alfani (2021), Ann Arbor, MI: Inter-university Consortium for Political and Social Research, 2021-02-26. https://doi.org/10.3886/E119067V1-81040

half of the variation of income in a global perspective—not accounting for ethnic origin or gender—is explained by two factors beyond individual control: country of residence and income distribution within that country.[5] These are not the only factors over which the individual exercises little or no control.

1.1 Explanations of Inequality

Ubiquity, tendency to grow over time, and context-dependence are three main characteristics of inequality that need to be captured by an explanation of the origins of inequality.

Tentative Explanations
A common explanation for differences in incomes or assets is that individuals differ with respect to *talent*, *effort*, and *patience*. A quick glance at historical, as well as contemporary, societies shows that this explanation, although clearly relevant in many situations, is inadequate. Differences in income or assets in classical Egypt or medieval Europe were far larger than can be explained in this way. A modern example is that the inequality with respect to numeracy and literacy skills globally is much smaller than the inequality in earnings.[6] Skill inequality explains only about 7 per cent of cross-country differences in earnings in advanced countries.[7] A further

---

[5] Milanovic (2015).

[6] Castelló and Doménech (2002), Green et al. (2015). Inequality is measured by the Gini coefficient.

[7] Devroye and Freeman (2001).

problem with the talent- or effort-based approach is that the secular drive towards increasing inequality appears difficult to explain. The social dependence illustrated in the example above is also absent.

An alternative explanation for differences is *violence* or threats of violence. European colonisation of the African and American continents was indeed based on violence, and the same can be said about Muslim expansion in the Middle East and North Africa. The colonial period has left an imprint on the societies affected, even after decolonization.[8] On the other hand, violence is a resource-consuming activity. A threat of violence is more efficient, but it must be translated into action at regular intervals in order to remain credible. Increasing inequality in most of the OECD countries in the last three to four decades is certainly not the result of increased use of violence. Internationally, the general level of violence within and between societies has been stable or decreasing for a long time.[9]

Although the above-mentioned factors can contribute to the understanding of how differences in assets and power originate and are maintained, something more is obviously required in order to explain the general pattern of increasing differences across time and space.

An Excursion: The Origin of Oscillations
As an illustration of the general approach taken in this book, a different but related problem will be discussed first—the origin of oscillations. Periodic motion, or oscillations, appear in many different natural and social contexts—heartbeats, daily or seasonal temperature variations, business cycles and many more. Some of these phenomena are easy to understand as effects of external influences, whereas others are more complex. Temperature variations across a 24-hour time span are direct consequences of variations in the influx of solar energy that are generated by the rotation of the Earth around its own axis. The Earth revolves around the Sun in a year, which in combination with the angle between the Earth's axis and the orbital plane gives rise to seasons, more or less pronounced depending on location. On a much longer time scale, small variations in the orbital regime of the Earth lie behind climate variations registered as ice ages. These mechanisms have been more difficult to analyse, but consensus on the so-called Milankovitch cycles now prevails.[10]

Nature offers more complex examples, however, where oscillatory motion is the *emergent result* of interaction between the components of a system or between the system and its environment. Consider a reed growing in a stream. If the stream flows slowly, the reed will move gently with small variations in the current. If the current becomes stronger, the reed may bend somewhat more but will otherwise not behave differently. If the current becomes sufficiently strong, however, the reed will start wobbling, swinging back and forth with a regular frequency in a plane that is perpendicular to the current. The reason is that vortices will be formed downstream

---

[8] Acemoglu and Robinson (2006, 2012).

[9] For illustrations, see, for example, Klingenstein et al. (2014) (UK, late eighteenth through early twentieth century), and Gleditsch et al. (2002) (post-WW-II era).

[10] See, for example, Imbrie and Imbrie (1979) or Crowley and North (1991).

of the reed, asymmetrically and alternating to the left and to the right of the centre line.[11] These vortices will draw the reed back and forth with a frequency that depends on the strength of the current. If the current becomes even stronger, transition to turbulence will occur, and periodicity will be lost.

A similar mechanism is active in a category of wind instruments, appropriately called *reed instruments*, represented by the clarinet and the saxophone. A tone is generated by forcing the reed attached to the mouthpiece to vibrate, and the tone is then amplified in the tube that forms the main part of the instrument.

It is important to see the difference between the periodic temperature variations described above and this motion. In the former case, the external force acting on the system is periodic, and the periodicity of the resulting motion is no cause for surprise, even if secondary effects such as ice ages involve a number of processes in the geosphere. By contrast, the force from the current acting on the reed is constant, and the oscillating movement emerges as a result of a complex interaction between the water or the air, respectively, and the reed.

Under certain circumstances, nature exhibits oscillations in aggregates even in the absence of such external forces. In the Mediterranean fishing industry, large swings in the catches of different species have long been observed. A few decades into the twentieth century, statistician Alfred Lotka and mathematician Vito Volterra engaged in solving this problem and independently developed a mathematical theory that showed the possibility of sustained regular oscillations in the populations of predator and prey species.[12] Similar oscillations have been observed in the deliveries of furs from the snowshoe hare and the lynx to the Hudson Bay company, and unless this reflects some strange periodic behaviour among trappers, it can be interpreted as an indicator of periodic swings in these populations.[13] Theory has developed since Lotka and Volterra published their original works, and the situation is more complicated than can be rendered by a simple two-species model, but the basic conclusion remains intact—interaction between predator and prey populations can generate sustained oscillations even in the absence of external factors acting on them.

Business cycles have been an important research topic among economists, and it was realised early on that nonlinear effects in the interplay between supply and demand could lead to sustained periodic variations in macro-variables such as output and employment.[14]

These examples of oscillatory behaviour can be categorised with respect to the degree of autonomy that the oscillating system exhibits with respect to the environment, as follows:

- *Oscillations driven by the environment*: The system is subject to periodic forces in the environment that engender periodic motion in the system with the same

---

[11] So-called Kármán vortices, named after the Hungarian-born physicist Theodore von Kármán. For an introduction, see Tritton (1988).

[12] Lotka (1925), Volterra (1926).

[13] Keith (1963).

[14] Kaldor (1940) is a classic reference. For a short survey of business cycle models, see Skott (2013).

frequency, possibly with secondary effects that follow more complex patterns (daily, seasonal, and secular temperature variations).

- *Oscillations driven by the interaction between the environment and the system*: The system is subject to a nonperiodic external force that gives rise to periodic motion as a result of the interaction between the system and its environment (reed vibrations).
- *Oscillations internally generated*: The system exhibits periodic oscillations that are generated internally (variations in population sizes, business cycles).

The Origins of Inequality: A Taxonomy

Analogously with the categorisation of oscillatory behaviours presented above, three different types of inequality will be identified, using the same defining parameter, that is, the degree of autonomy vis-à-vis the environment. The discussion will be based on what is one of the simplest possible types of social interaction—a game of marbles with two players. Equality or inequality in this case naturally refers to the number of marbles possessed by each player. For simplicity, it will be assumed that each round is played with one marble each and that the winner takes these two marbles. In practical terms, the winner could be the player that comes closest to a stick that has been fixed to the ground. The rules of the game and the general conditions may vary. The players may be equally skilful or differ in their skill level. They may start with the same number of marbles or with different amounts. A natural stopping rule would be that one of the players runs out of marbles, but whether the other player keeps the marbles or returns them is of course a matter of choice. Because there is an element of chance in this game, outcomes refer to expected values.

The appropriate mathematical tool of analysis for this game is the Markov chain.[15] Some conclusions from the analysis can be summarised as follows:

- If the players are equally skilful and start with the same number of marbles, they stand an equal chance of winning (which is intuitively clear from symmetry).
- If the players are equally skilful but start with different amounts of marbles, the probability of winning is proportional to their respective number of marbles.[16]
- A player that starts with fewer marbles than the other player can compensate for this with higher skill.

Against the backdrop of these results, it is possible to sketch three situations that correspond to the list of the preceding section. Assume first that the players are equally skilful but possess different amounts of marbles at the outset. The prehistory of this asymmetry could be that the richer player has inherited marbles from a sibling that has developed other interests, or that wealthy parents have provided the means for buying marbles. Another possibility is that this surplus was generated in earlier

---

[15] See Sect. 2.6 for the mathematical details.

[16] That is, if players 1 and 2 start out with n_1 and n_2 marbles, respectively, the probability of player 1 winning is $n_1/(n_1 + n_2)$, and analogously for player 2.

games with other players. From point 2 above it can be concluded that the richer player is likely to win, that is, the outcome is determined by the environment, in this case the initial holdings. The initial asymmetry of the distribution is further reinforced by the game.

Next, assume that the initial holdings of the players are equal but that one of them is more skilful than the other. In this case, the more skilful player will win. The outcome emerges from this asymmetry of skills and is, in this sense, internally generated in the game.

Intermediate cases are possible, in which initial holdings differ, but this difference is to some extent balanced by greater skill in the disadvantaged player. The threshold where the difference in skills balances the difference in initial holdings can be computed from the parameters of the problem. For instance, if player 1 starts with 5 marbles and player 2 with 50, player 1 needs to be approximately 15 per cent more skilful than player 2 in order to stand an equal chance of winning.[17]

In order to conceive a situation where inequality is completely internally generated, assume that the players are equally skilful, and that they possess an equal number of marbles at the outset. Assume further that a sequence of games is played, in which each round follows the same rules as before. At the end of each round, the winner keeps the marbles from that round. Assume for concreteness that they start with 5 marbles each and that they receive another 5 marbles each before the next round begins. Under these conditions, the outcome of the first round is open; both players stand a 50-per-cent chance of winning. As the second round starts, the winner of the first round has 15 marbles and the other player 5, so it is highly likely that the former will win also the second round (with 75 per cent probability). With each round, the distribution becomes more asymmetrical. From a completely symmetric starting condition, a strongly asymmetrical distribution has developed.

The results are summarised in the list below, with the three main cases corresponding to the three main categories of oscillatory behaviour presented above.

- *Inequality driven by the environment*: Holdings are asymmetric at the start, skills are equal, and asymmetry is reinforced by the game.
- *Inequality partly driven by the environment, partly internal*: The initial distribution is symmetric or asymmetric, skills are different, and the asymmetric outcome distribution depends on both assets and skills.
- *Inequality generated internally*: Symmetric initial conditions prevail with respect to skills and holdings, repeated games and symmetric incomes. An asymmetric outcome is stochastically generated by the game.

No analogy is perfect, and the analogy between oscillations and inequality as illustrated by the above examples is no exception to this rule. The taxonomy sketched may nonetheless serve the purpose of bringing some order into both the abundant empirical literature on inequalities and the more normative ethical and political literature. For all its simplicity, this set-up evokes some of the generic

---

[17] See Sect. 2.6.

problems of inequality—the importance of initial conditions and the sometimes intricate interplay between the environment and the system, as well as the possibility of inequality that is spontaneously generated from symmetric initial conditions— what is referred to as *symmetry breaking*.

It should be stressed that the categories used so far are purely qualitative. Nothing has been said on the issue of whether differences in outcome are small or large or reasonable with respect to differences in initial conditions or skill.

Static Versus Dynamic Explanations
In summary, the main idea to be developed in this book is that a theory of inequality has to be dynamic in order to successfully explain the ubiquity and historical patterns of inequality. A recurrent mechanism of high importance is the inherent instability of an equal distribution of resources that stems from social interaction between individuals or between groups, requiring neither difference between individuals nor the use of violence for inequality to develop. Instabilities of egalitarian distributions at the micro level provide a key to understanding why inequality arises and grows over time.

1.2 Plan of the Book

Mathematical models play an important role in this book, but in order to increase accessibility for readers who are less familiar with such tools, the exposition is separated into descriptive chapters and chapters where formal mathematical reasoning is used.[18] Chapters 3, 5 and 7 plus the concluding two chapters could in principle be read in a sequence by readers who are content with an intuitive and empirically oriented idea of the main results and arguments. More precise statements and sketches of proofs are presented in the Chaps. 4, 6 and 8. These chapters, as well as two technical sections of Chap. 2, are marked with an asterisk. The mathematical background needed for these chapters comprises general analysis, differential and difference equations (in one case, stochastic differential equations), and stability. For full proofs, the reader is referred to background publications. The notation mainly follows that of the background literature, which implies that capital is sometimes denoted by K, in other cases by Θ, and so on. The summaries at the end of Chaps. 4, 6 and 8 are written in non-technical language and are also recommended to readers who are interested mainly in qualitative results.

Chapter 2 contains a succinct presentation of the main analytical tools to be used in later chapters. As should already be clear from this introductory chapter, concepts such as dynamical systems, instability, games and negotiations play a central role, as well as—of course—definitions and measurement of inequality. The presentation is deliberately kept at a terse minimum. There is an abundance of basic textbooks on

---

[18] Readers will recognise this design from Sen (1970/2017).

inequality, dynamical systems, game theory, and so on. The advantages and risks associated with mathematical models of different designs are discussed. This chapter also contains the mathematical background to the paradigmatic examples of oscillations and the game of marbles used for illustration above.

Chapter 3 is devoted to inequality from a life cycle perspective. The focus is on mechanisms that tend to generate (or possibly dampen) inequality, from conception through to old-age retirement. The perspective is primarily individual, but even so the mechanisms and processes used for illustration of course presuppose some interaction with the environment and often with other human beings. One of the topics covered is the classical nature-nurture debate and the new light that has been shed on this subject of contention by research on epigenetics. A growing literature focuses on the early stages of individual development—prenatal, perinatal and early years—what is referred to as *the foetal origins hypothesis*. The environment, broadly defined, may intervene in two ways, either by providing or not providing enough stimulus, or by actually harming the individual. Such deviations from normality are not uniformly distributed over a population and are consequently a source of inequality.

Education is fundamental to a child's preparation for adult life, but recurrent education is increasingly also an important factor in working-life. Health is another fundamental dimension of individual welfare that is unevenly distributed across the population—like education, with which it also co-varies. Early development, education and health are all areas where public interventions are considered appropriate, and there is also a growing literature on the evaluation of such interventions.

The following chapter describes some of the main approaches to the measurement and modelling of individual development. The multitude of factors that affect human development create a problem in identifying causal relationships in this field. Between education and health, for instance, causality goes in both directions. Models in this field often rely on some capital concept—human capital for knowledge and networks, health capital for health status—and the way in which the mechanisms of capital accumulation and reproduction are rendered by the models is of key importance.

The transfer of capital between generations is also an important factor in explaining the persistence and growth of inequality. This concept refers not only to the capital forms cited above but also to physical assets and financial capital.

Chapter 5 changes the perspective from the individual to the social level. Starting from binary and small-group interaction, inequality-generating mechanisms are traced through networks and into fully developed markets. As in individual development, there is an environment that intervenes in different ways, but the focus is on the interaction itself and how inequality emerges under varying assumptions about the interacting parties.

Chapter 6 provides some mathematical tools for understanding these mechanisms. Interaction at the two extremes—bilateral interaction on the one hand and large populations with fully fledged markets on the other—is in general mathematically more tractable than intermediate group sizes.

Chapter 7 is devoted to spatial inequality. Human activities are not uniformly distributed across the available territory, and differences visible in the size distribution of agglomerations can be explored using perspectives and concepts that are similar to those used for the analysis of individual inequalities. Arguments about fairness or justice are seldom raised directly in connection with these differences, but differences in opportunities, incomes and assets emerge indirectly as a result of spatial heterogeneity, so there exists a link to classical problems of distribution. The subsequent chapter elaborates some of the main mathematical models developed for the analysis of spatial allocation processes.

In Chap. 9, some of the main conclusions from previous chapters are pulled together as a basis for an ethical and policy-oriented discussion on inequality. The natural point of departure is that the legitimacy of a certain distribution of incomes or assets in a society is intimately connected to its origin, and that the insights gathered from previous chapters may be used to navigate among the plethora of moral and political arguments used in this field.

The final chapter widens the perspective to the role of analysis in political economy and discusses some possible paths for theoretical development.

Bibliography

Acemoglu, D., & Robinson, J. (2006). *Economic origins of dictatorship and democracy.* Cambridge University Press.

Acemoglu, D., & Robinson, J. (2012). *Why nations fail. The origins of power, prosperity and poverty.* Profile Books.

Alfani, G. (2021). Economic inequality in preindustrial times: Europe and beyond. *Journal of Economic Literature, 59*(1), 3–44.

Bourguignon, F., & Morrisson, C. (2002). Inequality among world citizens: 1820–1992. *American Economic Review, 92*(4), 727–744.

Bowles, S., et al. (2010). Intergenerational wealth transmission and inequality in premodern societies. Special issue of *Current Anthropology, 51*(1).

Castelló, A., & Doménech, R. (2002). Human capital inequality and economic growth: Some new evidence. *The Economic Journal, 112,* C187–C200.

Crowley, T. J., & North, G. R. (1991). *Paleoclimatology.* Oxford University Press.

de Waal, F. (2005). *Chimpanzee politics. Power and sex among Apes. 25th Anniversary edition of the original 1980 book.* Johns Hopkins University Press.

Devroye, D., & Freeman, R. B. (2001). Does inequality in skills explain inequality in earnings across advanced countries? *NBER working paper* No. 8140. Cambridge, MA: National Bureau of Economic Research.

Gleditsch, N. P., et al. (2002). Armed conflict 1946–2001: A new dataset. *Journal of Peace Research, 39*(5), 615–637.

Green, A., et al. (2015). Cross-country variation in adult skills inequality: Why are skill levels and opportunities so unequal in anglophone countries? *Comparative Education Review, 59*(4), 595–618.

Imbrie, J., & Imbrie, K. P. (1979). *Ice ages. Solving the mystery.* Harvard University Press.

Kaldor, N. (1940). A model of the trade cycle. *Economic Journal, 50,* 78–92.

Keith, L. B. (1963). *Wildlife's ten-year cycle.* University of Wisconsin Press.

Klingenstein, S., et al. (2014). The civilizing process in London's Old Bailey. *Proceedings of the National Academy of Sciences, 111*(26), 9419–9424.

Lindert, P. H. (2000). Three centuries of inequality in Britain and America. Ch. 3 in A. B. Atkinson, & F. Bourguignon (Eds.), *Handbook of income distribution*, Vol. 1. Amsterdam: North-Holland

Lotka, A. J. (1925). *Elements of physical biology*. Williams and Wilkins.

Milanovic, B. (2015). Global inequality of opportunity: How much of our income is determined by where we live? *The Review of Economics and Statistics, 97*(2), 452–460.

Morrison, C. (2000). Historical perspectives on income distribution: The case of Europe. Ch. 4 in A. B. Atkinson & F. Bourguignon (Eds.), *Handbook of income distribution*, Vol. 1. Amsterdam: North-Holland.

Scheidel, W. (2017). *The great leveler. Violence and the history of inequality from the stone age to the twenty-first century*. Princeton University Press.

Sen, A. (1970/2017). *Collective choice and social welfare*. San Francisco: Holden-Day (1970). Expanded ed. Cambridge, MA: Harvard University Press.

Skott, P. (2013). Business cycles. In J. E. King (Ed.), *The Elgar companion to post Keynesian economics* (2nd ed.). Edward Elgar Publishing.

Tritton, S. (1988). *Physical fluid dynamics* (2nd ed.). Oxford University Press.

Volterra, V. (1926). Fluctuations in the abundance of a species considered mathematically. *Nature, 118*, 558–560.

Chapter 2
Preliminaries

Inequality is a neutral concept, unlike *inequity*, *fairness* and *justice*, which are used in philosophical literature. Not all inequalities give rise to fairness discussions. For instance, a classical observation in economic geography is that the size distribution of cities exhibits a fairly stable pattern across time and space,[1] but few if any would argue that it is unfair that the biggest city is that much larger than the second-largest, and so on. The main purpose of this book is to describe mechanisms that tend to generate inequality, so the aim is descriptive rather than normative. This is not to say that description or analysis is irrelevant to discussions on equity or fairness. On the contrary, an in-depth understanding of social processes is crucial to the development of coherent and tenable theories of fairness, and it is surprising how much has been written on these matters by political philosophers without anchoring in empirical social science.[2]

2.1 Some General Remarks on Mathematical Models

The term *model* is used in different, more or less formalised, senses. A person riding a bicycle has a *mental* model of the dynamics of the bicycle and the rider, and uses this model in order to manoeuvre the vehicle in the desired direction, in order to avoid obstacles, and so on. Human decision-making is often guided by *verbal* models or thumb rules of the type "if the interest rate goes up, stock prices will go down". This is a different kind; the mental model of the bicycle rider would be difficult to verbalise. A third kind would be physical or *iconic* models, that is, scaled-down versions of the system to be analysed, sometimes used for engineering purposes.

---

[1] See Chap. 7 and references therein.

[2] Normative implications are discussed in Chap. 9.

© The Author(s), under exclusive license to Springer Nature Switzerland AG 2022
P. Molander, *The Origins of Inequality*,
https://doi.org/10.1007/978-3-030-93189-6_2

In the present context, the main concern will be with *mathematical* models. Such a model requires that a *system* be delineated from the environment and that a number of variables be selected for the description and analysis of the system. The aim of the analysis could be simply to acquire knowledge about the system but also to influence it with some purpose in mind. Whatever the ultimate purpose, the modelling and analysis require a number of critical steps:

- *Delineation of the system*: A borderline must be drawn between the system and the environment, the latter comprising factors that are considered as "given" or as "disturbances".
- *Creation of the model*: This step presupposes a *mapping* from the real system into a set of variables forming the *state* of the system.
- *Analysis*: During this phase, the strength of mathematical and statistical tools is mobilised in order to explore the characteristics of the system and the consequences of changes in assumptions, as well as the effect of different policy alternatives when relevant.
- *Implementation*: The conclusions from the mathematical analysis have to be translated back into the real world. Mathematical statements are of limited use unless they can be re-shaped into meaningful statements about the system modelled or into implementable policy advice, if this has been the purpose of the model.

The above steps represent different kinds of difficulties. Analysts with a background in the mathematical or statistical sciences have a large toolbox and are well versed in the manipulation of models. Such topics are taught in textbooks, and academic literature constantly supplies new and more sophisticated variations of established model design and analysis tools. For this part of the problem, there exists a technology in the form of standardised models and techniques of analysis.

The modelling and implementation phases, by contrast, could be characterised more as an art than as a technology. Although there is an abundance of textbooks on modelling in the social sciences, it takes a long period of practice to acquire expertise in the choice of models, in finding the right trade-off between realism and manageability, and in interpreting the mathematical conclusions in the real-life context for which the model was developed.

The situation resembles that of an orienteering athlete, for whom the task is to find the best way from point A to point B in a dense and boggy forest. The fastest route is most often not the straight line from A to B. If there is a path or a road on the way, the best course of action may be to aim for that path during part of the course. The problem is to find the right points where to join the path and where to leave it.[3] In social science, mathematics is that path to swift conclusions. The problem is to know the best points where to join and where to depart.

---

[3] In idealised situations, Snell's law in optics will give an indication. But real-life situations are often far from the ideal.

A common error is to carry a preconceived idea of which type of model is appropriate for a given problem, for traditional reasons, or because that is where the analyst has his or her expertise and feels at home. By way of example, linear models—models that can be scaled up or down without altering their qualitative behaviour—dominate as a result of their relative mathematical simplicity, but it must be recognised that the very choice of a linear model excludes certain types of behaviour, which will consequently be filtered out from the analysis, in many cases unconsciously. Both linear and nonlinear models can be used for generating oscillations, but linear models are sensitive to perturbations of both initial conditions and parameters. Changing the initial condition of a linear system will normally lead to a different amplitude, and even a small change of a parameter value can lead to a qualitatively different behaviour, such as convergence to a stable equilibrium or instability. By contrast, nonlinear models may give rise to limit cycles, which are insensitive to small disturbances of both initial conditions and parameter values.[4] A system that exhibits this type of behaviour consequently requires a nonlinear model.

Another type of error is to underestimate the need for first-hand knowledge about the system. Among analysts who have developed high skills in mathematical and statistical tools, there is a temptation to try to compensate for inadequate knowledge about the system with more sophisticated statistical methods. This will most often lead the analysis astray. There are important types of error that cannot be eliminated in this way.

In the social sciences, any model, however sophisticated, is an approximation.[5] There are always important aspects that have been neglected, and conclusions should be presented with an appropriately humble attitude. Neglected aspects may be important for the conclusions drawn. Policy recommendations based on a model assuming full information, for instance, will tend to favour persons who are well informed at the expense of those who are less well informed.

2.2 Inequality: Basic Concepts

The academic literature on inequality offers a wide variety of concepts and tools, developed both for descriptive and for normative purposes.[6] Basic components of the toolbox include *unit of analysis*, *focus variable*, *measure of inequality* and *time perspective*.

---

[4] See Sect. 2.5 for more details.

[5] In physics, there is a discussion as to whether models are approximations or whether mathematics in a more profound sense represents the deeper nature of reality; for a representative of the latter view, see Tegmark (2014). That discussion need not concern us here.

[6] Overviews can be found in, for example, Atkinson (1983), Cowell (2000), or Jenkins and Van Kerm (2009).

Unit of Analysis

Depending on the application, the appropriate unit of analysis may be individuals, households, groups, enterprises, cities, or other aggregates. Knowledge and health naturally refer to individuals. Incomes or assets often refer to individuals, but in the presence of transfers between individuals belonging to the same household, the latter becomes a more appropriate unit. In this case, all individuals belonging to the same household are assumed to enjoy the same standard of living, which may underestimate the actual level of inequality. If the purpose of the analysis is to describe the differences between women and men, their incomes must of course be kept apart, which may by contrast lead to an overestimation of actual differences. Other categorisations may be interesting, based for instance on social strata or geography.

Focus Variables

Given an individual focus, variables of interest may be characteristics such as body height, IQ test values, income, assets, or social status using some recognised status scale. Typical questions asked would include, for instance, how these variables correlate and what external factors affect them.

When households are in focus, standard of living rather than pure income becomes the natural indicator, and assumptions have to be made on transfers within the household and costs of living. The normal course of action is to render returns to scale in living costs using a square-root function or some empirical estimate, and to assume identical levels for all members of the household (which normally, as noted above, underestimates the level of inequality).

For other aggregates such as enterprises or cities, turnover and population or local GDP, respectively, are commonly used focus variables.

In many situations, the target is multidimensional. Human well-being is a function of health, education, income and other variables, and it is not obvious if, and in that case how, they should be summarised. Attempts have been made to form a weighted index of the focus variables, such as the Human Development Index. This sometimes generates unacceptable results, however,[7] so care must be exercised when creating such composite measures. It may be preferable to keep the target variables apart.

Measures of Inequality

Social scientists use a wide range of measures of inequality—the coefficient of variation, 90/50 ratios, the Gini coefficient, the Theil measure and others—which summarise certain characteristics of the underlying distribution of incomes or assets.[8] The coefficient of variation is simply the ratio between the standard deviation (σ) and the average (μ). The 90/50 ratio is the ratio between the average income of the 90th percentile group and the median income. The 50/10 ratio is defined similarly. The *Gini coefficient* or *Gini index*, which is perhaps the most commonly used index, measures the distance between the actual distribution and a completely

---

[7] Ravallion (2012).

[8] For overviews, see note 6.

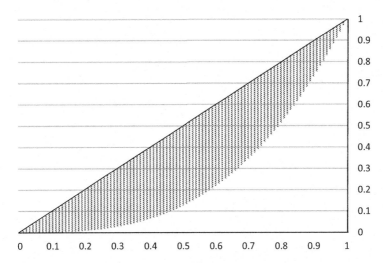

Fig. 2.1 Definition of the Gini coefficient. The coefficient is defined as the surface of the shaded area divided by the surface of the triangle

equal distribution. It is defined in the following way. Order the population p with respect to the focus variable, for instance income, starting with the lowest values. Normalise total population and total income to 1, so that the cumulative distribution function $L(p)$[9] for income varies from 0 to 1 over the interval [0,1]. The Gini coefficient G is then defined as twice the surface between the straight line that corresponds to full equality and the actual distribution, as illustrated in Fig. 2.1.

When all incomes are equal, $G = 0$. The other extreme is that the whole income is concentrated to one individual, in which case $G = 1$. The Gini coefficient is also a measure of the average difference between two individuals in the population for a given average; it can be shown that the expected value of this difference equals $2\mu G$, where μ is the average income.

When the focus is on groups or regions rather than individuals, the definition is similar. Values for individuals will be replaced by group averages, but otherwise the mathematical definitions are the same.

Whatever measure is used, information is lost in the transformation from the distribution to the measure used. Two societies having the same Gini coefficient can be very different when it comes to social structures or incentives. Consider, for instance, the two societies, A and B, shown in Fig. 2.2. In society A, 40 per cent of the population receive 1 per cent of the income and the remaining 60 per cent divide the rest equally. This could correspond to a situation where an aboriginal minority is economically exploited by an upper class of historic invaders. In society B, a large majority of 99 per cent is controlled by a small elite that receives 40 per cent of the total income. It is obvious that social relations and economic incentives in these two

---

[9]The function $L(\cdot)$ is referred to as the *Lorenz curve*.

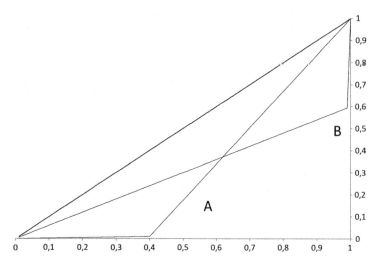

Fig. 2.2 Two societies with different income distributions having the same Gini coefficient

societies are radically different and that socioeconomic development cannot be expected to follow the same path, although the Gini coefficients are in fact equal.

If the focus is on *poverty*, lower educational or income strata are particularly interesting, so measures such as the ratio between the average income of the lowest decile group and a medium income (10/50 ratio), or between the lowest decile and the uppermost stratum (10/90 ratio), will be relevant measures.

All the measures of inequality mentioned are *relative*—they refer to shares of a given total. A 10 per cent increase in income for everyone in a society will leave the Gini coefficient unchanged, but such an increase will expand the economic room for manoeuvre more for high-income than for low-income earners.

Any scalar measure of an income distribution presupposes the compression of an entire distribution into one single number, and a lot of information is necessarily lost in that operation, as the above example about the Gini coefficient illustrates. When possible, it is therefore preferable to keep a more disaggregated picture of the distribution, based, for instance, on the average incomes of the decile or quintile groups.

Time Perspective
The measures described above refer to single points in time. In some contexts, the focus is on longer time periods or on changes between two points in time. If incomes for one and the same individual or household can be measured over a longer period, averaging is possible. Shorrocks' index[10] is intended to describe how income differences are reduced when incomes are averaged over a longer period than a

---

[10]Shorrocks (1978).

single year.[11] Because some volatility is filtered out this way, inequality tends to be reduced. The overall effect depends on the profile of volatility across the population.

Longer periods may also refer to probabilities of change for one and the same individual, or to links between generations. Such mobility—of incomes, education, or some other status variable—can be represented by scatter plots for the distribution or the joint distribution function at the two points in time that are studied. Commonly used measures of mobility are based on correlations between the two points in time or between generations.

2.3 Dynamical Systems

The mathematical theory of dynamical systems has been developed for the purpose of analysing how systems develop over time, either at discrete points in time or continuously, using difference and differential equations, respectively.

State Variables, Controls, and Feedback
The current status of a system is described using an n-dimensional state variable $\vec{x}$ that varies with time t: $\vec{x}(t) = (x_1(t), x_2(t), \ldots, x_n(t))$. In some applications, for instance when the economic activities across a territory are analysed, the state vector will have a very high dimension, and in some applications, it is customarily assumed that the underlying space is infinite-dimensional. The time development of the state variable $\vec{x}(t)$ is governed by a differential equation (continuous time) or a difference equation (discrete time):

$$\frac{d\vec{x}}{dt} = f\left(\vec{x}(t), \vec{u}(t)\right) \ (continuous \ time)$$

$$\vec{x}(t+1) = f\left(\vec{x}(t), \vec{u}(t)\right) \ (discrete \ time)$$

Here, $\vec{u}(t)$ is a vector of control or policy variables that can be used for influencing the system. If $\vec{u}(t)$ is absent, the system is autonomous. In engineering contexts, control variables are often defined as functions of the current state, $\vec{u}(t) = g(\vec{x}(t))$. This is the principle of *feedback*. The concept of feedback is also often used metaphorically in contexts such as biological or social systems, even in the absence of any visible engineer. Once a feedback law has been decided and put in place, the system becomes autonomous.[12]

---

[11] Björklund (1993), Burkhauser and Poupore (1997).

[12] A self-contained text on differential equations that includes the basics of control theory is Logemann and Ryan (2014).

Equilibrium and Stability

Central to this theory is the concept of equilibrium, basically a point where the system may come to rest. Given that real-life systems are always subject to perturbations, the behaviour of the system close to equilibrium points is of fundamental importance. An intuitive idea of stability of an equilibrium is that the system will return towards the equilibrium following a small disturbance. The appropriate mathematical concept is *asymptotic stability*; an equilibrium is asymptotically stable if, when starting sufficiently close to this equilibrium, the system will remain close to it and return to the equilibrium as time goes to infinity.

It has often been assumed that only stable equilibria are worthy of interest. When general systems theory developed in the 1950s, the concept of *homeostasis* was used to designate mechanisms in any system that operate to keep it within a close range of a desired point, such as a body temperature of 37 °C.[13] In economic theory, it has similarly been tacitly assumed that the general equilibrium around which mainstream theory revolves is stable. The existence of business cycles would seem to contradict this idea, but they have been considered to be acceptable deviations from a basically stable equilibrium. Nonetheless, the existence of unstable equilibria in the classical model has been shown theoretically by several authors.[14]

Contrary to this focus on stable equilibria, it will be argued in what follows that self-reinforcing mechanisms and instability are fundamental for the emergence of inequality in social systems. When subject to a disturbance, a system may respond by dampening the disturbance (negative feedback) or by reinforcing it (positive feedback). Negative feedback is a necessary (but not sufficient) condition for stability, so positive feedback will lead to instability. Variants of these phenomena are known under names such as *self-reinforcement*, *vicious circles* (colloquially) or *returns to scale* (in economic language).

2.4   Games

A defining characteristic of interaction in human societies is that everybody affects the outcome, but nobody is in full control. Basic to any rationalistic approach to human behaviour is that human beings have choices to make in everyday life. For the term *choice* to be meaningful, there must be more than one alternative. When combined with the choices of fellow human beings, choices lead to *outcomes*. Outcomes are ranked according to some reference scale of good and bad. Alternatives may refer to more or less anything—having a certain amount of food available, being married to X or not, etcetera. Most of the outcomes that will be in focus in the discussion that follows come in quantities and refer to resources of various kinds—education, health, physical or financial assets, and so on.

---

[13] For further examples, see Emery (1969).

[14] An early example is Scarf (1960); see further Sonnenschein (1972) and Mantel (1974).

The mathematical theory developed for the analysis of social choice situations is *game theory*.[15] The term *game* is somewhat unfortunate; we tend to associate it with children, play and pastime activities, but game theory can be applied to serious problems. Further, when adults are involved, we tend to think of games as connected with profound thinking and strategic behaviour, as in a game of chess, but such an assumption is not necessary for the theory to be applicable, at least not in its modern form. Social experimentation, where new modes of behaviour are born, tested by trial and error and spread, is by no means inaccessible to the theory.

The simplest society that can be imagined consists of two persons, *1* and *2*. If player *1* has two alternatives at his disposal and player *2* has three, this yields six combinations in all, assuming that they can choose independently. The way in which the players rank these six possible outcomes will determine the dynamics of their social interaction. This is the simple idea behind game theory.

Ranking may be *ordinal* or *cardinal*. In the former case, the players simply order the outcomes according to their preferences. In the latter case, players are assumed to have a *utility scale u(·)*, which measures how much a given alternative is preferred to another. The utility function *u* is assumed to be increasing (more is better) and concave (saturation).[16]

A combination of choices from which no party unilaterally has an incentive to deviate is called a *Nash equilibrium*. An alternative such as the ones open to players *1* and *2* above are called *pure strategies*. In some games, the best choice is not such a pure strategy but turns out to be a probabilistic combination of alternatives—for instance that one alternative should be chosen with a probability of 0.4 and the other with a probability of 0.6. This is called a *mixed strategy*. If such combinations are accepted as solutions, it can be shown that every game has at least one equilibrium.[17]

An example of a game represented by its payoff bi-matrix is shown in Fig. 2.3.

The Three Archetypes
The computational complexity of games increases rapidly with the number of agents and alternatives. Even in a minimal set-up, with two players who have two alternatives each, there will be four outcomes. These outcomes can be ranked in 24 different ways by each of the players independently, which yields $24^2 = 576$ different possibilities. Many of these combinations will in fact be identical, however, resulting from a relabelling of players and alternatives. There are 78 genuinely different plays with two players and two alternatives each.[18]

Under additional assumptions, the list of alternatives can be further condensed. A game is *symmetric*, if both players have the same set of strategies and the payoffs are independent of the labelling of the players, that is, they play under equal conditions ($s_{ik} = s_{jk}$ and $a_{ij} = b_{ji}$ for all *i,j* in the bi-matrix in Fig. 2.3). Further, the ranking of

---

[15] A basic reference for classical game theory is Fudenberg and Tirole (1991).

[16] In mathematical terms, $u'(x) > 0$ and $u''(x) < 0$ for all x in the domain of $u(·)$.

[17] This was one of the main achievements of John Nash in his PhD dissertation, published in 1950 (Nash 1950a), reprinted in Kuhn and Nasar (2002).

[18] Rapoport and Guyer (1966).

Fig. 2.3 A game with two
players who each have two
alternatives. Player A has
alternatives s_{11} and s_{12}, and
player B has alternatives s_{21}
and s_{22}. Payoffs are denoted
by a_{ij} and b_{ij}, respectively

	S_{21}	S_{22}
S_{11}	a_{11}, b_{11}	a_{12}, b_{12}
S_{12}	a_{21}, b_{21}	a_{22}, b_{22}

Fig. 2.4 The reduced form
of symmetric 2×2 games

	S_1	S_2
S_1	a_1	0
S_2	0	a_2

two alternatives is unchanged, if the same amount is subtracted from both payoffs. This can be used to reduce the number of games by setting all off-diagonal elements equal to 0. In symmetric games, it suffices to have single numbers as entries, since payoffs to the players are equal. This yields the reduced matrix in Fig. 2.4.

Three alternatives are possible.

- *Both a_1 and a_2 are positive.* This is the so-called *coordination game.* Both players aim at choosing the same strategy, and preferably the one with the highest payoff, if there is a difference. A concrete example may be to drive on the left-hand or right-hand side of the road. Another concrete example is the *stag hunt*, once used by Rousseau in his *Discourse on Inequality* to illustrate the problems of coordinating social efforts in the hunt for game.[19] The players can hunt small game separately, but they can do better if they coordinate and go for the large game, the stag. If one player chooses the stag alternative but the other does not, the former gets nothing, so there is a dilemma of trust involved. Both s_1 and s_2 are pure Nash equilibria. Somewhat surprisingly, there is also a mixed Nash equilibrium,

---

[19] For a discussion, see, for example, Skyrms (2001).

corresponding to a probabilistic mixture of the two strategies. This equilibrium yields a lower payoff (think of the traffic application).

- *The payoffs have different signs.* Suppose, without loss of generality, that a_1 is negative and a_2 positive. Then s_2 is the dominant strategy—it is better than the other strategy irrespective of the other player's choice, and it is the only Nash equilibrium. The payoff to both players is a_2, which may be worse than a_1, the outcome if both players had chosen strategy s_1. This happens when $a_{11} > a_{22}$ in the original bimatrix in Fig. 2.3. This is the so-called *prisoners' dilemma*, a game that renders the structure of a large number of conflicts between individual and collective rationality—moral problems of everyday life, production of public goods, environmental degradation, security policy and others. It may be individually rational to lie, to steal, not to contribute to a public good, but if everyone follows this indication, the collective ends up with a negative outcome. Note that the outcome is negative following the evaluation of the players; there is no collective preference assigned from outside.
- *Both a_1 and a_2 are negative.* This is an *anti-coordination game*, where the aim of the players is to avoid the strategy chosen by the other player. It is a simple model of conflict, and a common name is the *hawk-dove game*, where *hawk* signifies aggressive behaviour and *dove* cooperative. Both pure strategies are Nash equilibria, but the most interesting Nash equilibrium is a mixed one, where both players choose to be hawks or doves with some probability depending on the parameters of the problem, thus minimising the damage.

The above format comprises games that are played only once. In real life, players often belong to the same community and expect to meet again; games are *repeated*. This opens the possibility of strategies that are contingent on the other player's choices, and a rich variety of new equilibria become possible. For instance, the prisoners' dilemma in repeated form has equilibria with a higher payoff under certain conditions.

Evolutionary Game Theory

Obviously, games in the form presented above become analytically intractable even for moderately complex situations. Assuming that players rank all the outcomes, that they are fully aware of the rankings of other players and that they are able to analyse the strategic interaction with other players is not realistic. *Evolutionary game theory* has been developed partly in response to this insight. In the evolutionary version, it is assumed that boundedly rational players try different alternatives within a given set of feasible strategies, and that successful strategies will multiply and spread in the population. Bounded rationality can have different meanings—information is incomplete or is not used even though it is available, players are myopic even if they expect the game to be repeated, and so on.

The evolutionary approach has a biological ring, and this was also the background, when it was introduced in the 1970s by John Maynard Smith and George

Price.[20] No foresight or strategic thinking is needed for this approach to work, only a mechanism that translates success in the interaction process into a survival rate. In this approach, the actual payoff to the players, and not just the ranking of outcomes, will matter. Payoffs will affect survival rates and the general pattern of behaviour.

An *evolutionarily stable strategy* (ESS) is a strategy that cannot be invaded by other strategies. It can be shown that an evolutionarily stable strategy corresponds to a Nash equilibrium, but the converse need not hold (that is, there are Nash equilibria that can be invaded by other strategies).[21]

Selection processes—much like biological processes—may operate quite slowly, and a population may get stuck in an equilibrium that is far from efficient. Natural selection is myopic, and models simulating natural selection run the risk of underestimating the capacity among human beings for seeing beyond the immediate vicinity of the status quo. Such myopia is nonetheless a fact in many contexts.

The transition from classical to evolutionary game theory opens up opportunities for the use of one of the most potent instruments of analysis that is available in mathematics—the theory of dynamical systems based on differential and difference equations. A natural approach is to define each strategy as a subpopulation and to assume that the net growth rate is proportional to the payoff that it receives in encounters with other strategies, the so-called *replicator dynamics*. In relative terms, the success of a strategy—its fitness—in relation to the average fitness will decide its growth rate.[22] In consequence, games that are difficult to analyse in their original form may become accessible to analysis by the transformation into an evolutionary framework and the associated differential equations.[23]

In evolutionary game theory, competition is between patterns of behaviour formalised as strategies. It is necessary to be precise about who or what is the *carrier* of this strategy. In biology, there is a long-standing debate on this issue. Various candidates have been advanced—groups, individuals, genes and even molecules.[24] The problem is no less difficult in the context of human societies.[25] Models of this type have been used for analysing a variety of social phenomena of relevance to the analysis of inequality—conflicts, the emergence of classes, and the birth of institutions.[26] Richard Dawkins, in *The Selfish Gene*, launched the concept of *memes* in analogy with genes:

---

[20] Maynard Smith and Price (1973). Maynard Smith (1982) provides a full exposition of the theory. For an update, see McNamara and Weissing (2010).

[21] See Weibull (1995), Chap. 2, for examples.

[22] Taylor and Jonker (1978).

[23] Weibull (1995), Chap. 3. For a comprehensive text also including stochastic models, see Sandholm (2010).

[24] The most common choice has been individuals. Wynne-Edwards (1962) argued for group selection. Dobzhansky (1958) argued for genes, which was taken up by Williams (1966) in a critique of Wynne-Edwards. Dawkins (1976) pursued this line further. Dover (1982) made a case for the molecular level. For an up-to-date discussion, see Jablonka and Lamb (2020).

[25] An early discussion can be found in Hirshleifer (1977).

[26] For examples, see Young (1998), Durlauf and Young (2001).

> Examples of memes are tunes, ideas, catch-phrases, clothes fashions, ways of making pots or of building arches. Just as genes propagate themselves in the gene pool by leaping from body to body via sperms or eggs, so memes propagate themselves in the meme pool by leaping from brain to brain via a process which, in the broad sense, can be called imitation.[27]

Obviously, so broad a concept is difficult to use in an analysis of social interaction.

A further difficulty when modelling interaction in social systems is that the mechanism of strategy updating must be described. Players that are dissatisfied with their current strategy may choose different rules for changing them—marginal change in some direction believed to improve the outcome, imitation of more successful strategies, and so on.[28]

A general risk when using an evolutionary framework is that the rationality of selection procedure is overstated. This risk must be acknowledged already in biological applications[29] and becomes all the more acute in the social sciences. The selection pressure against less successful strategies may be weak, particularly in a noisy environment. A condition for survival may simply be that strategies are, in Milo's words, "good enough".[30]

Stochastic Stability

Against the backdrop of the forceful evolutionary framework, it is tempting to see the concept of an evolutionarily stable strategy as the definitive answer to the question of which stability concept is most appropriate when analysing social interaction. One drawback is that disturbances in a system may alter its behaviour radically in the long run in such a way that the evolutionarily stable equilibria become less relevant. In order to cope with this problem, Foster and Young introduced the concept of *stochastic stability*.[31] Roughly speaking, a set S is a stochastically stable set if, in the long run, it is almost certain that the system motion is within an arbitrarily small neighbourhood of S as the noise tends to zero. If the aim of the mathematical model is to describe in what state the system can be expected to be found in the long run, this set is more appropriate than evolutionarily stable states. It is possible to find systems with a unique evolutionarily stable equilibrium but in which the motion of the system is in practice nowhere near this state; instead, it is found in the neighbourhood of the stochastically stable set.[32]

Multiple Populations

In many situations, both in nature and in human societies, encounters between the parties are not symmetric in the way assumed above. In nature, there are predators

---

[27] Dawkins (1976), Chap. 11.

[28] So-called *revision protocols*; Sandholm (2010) offers a detailed analysis of the consequences of different choices in this regard.

[29] Gould and Lewontin (1979), Milo (2019).

[30] Milo (2019). See also Hofbauer and Sandholm (2011).

[31] Foster and Young (1990).

[32] Ibid.

and prey, in human societies, producers and consumers, employers and employees, etcetera. This makes it necessary to imagine more than one population involved in the interaction process. In general, conclusions from the single-population setting will be modified.[33]

Remember that there are essentially three categories of symmetric 2×2 games: the coordination game, the prisoners' dilemma and the anti-coordination game. Among these, the prisoners' dilemma is unaltered by the transition to a bi-population setting. Prospects for cooperative behaviour are as bleak as before.

The conclusions from the analysis of the other two games are different, however. The mixed Nash equilibria turn out to be dynamically unstable in the multi-population setting, and the pure strategies are the only possible long-run outcomes. In the coordination game, this implies no loss to the players; getting rid of an inefficient potential outcome represents an improvement. In the hawk-dove game, by contrast, the loss of the mixed Nash equilibrium implies that the overall population is polarised into one subpopulation of hawks and another of doves. This is of obvious relevance to the perspective of inequality.

Negotiations

Bargaining and negotiation can be seen as a game. The term *bargaining* should not be understood too literally. It is simply assumed that two or possibly several parties have the possibility of concluding an agreement, that there is some conflict of interest about what agreement to conclude, and that no agreement can be imposed on any party without approval. In many situations, bargaining may be understood as a tacit understanding between two parties, without any formal bargaining whatsoever involved.

Much of the modern theory of bargaining is again related to the name of John Nash. Nash used two approaches in developing his theory. In the first, which he called *axiomatic*, he put together a set of reasonable basic requirements on the solution of the bargaining problem and proved that there is one and only one solution that satisfies these requirements.[34]

The second approach utilised by Nash—which is normally labelled *strategic*—was to assume a more formal bargaining situation, in which the parties bid and there is an agreement if the bids are compatible. Under certain additional assumptions, it can be shown that the axiomatic and the strategic approach will lead to the same result.[35]

Assume for simplicity that there are only two parties, *1* and *2*, that their respective levels of consumption are x_1 and x_2, and that the associated utility levels are given by the utility functions $u_1(x_1)$ and $u_2(x_2)$. Assume further that they will end up at the origin, (0,0), if no agreement is reached—what is called the *disagreement point*.

---

[33] Weibull (1995), Chap. 5.

[34] Nash (1950b, 1953).

[35] For a derivation, see Easley and Kleinberg (2010), Chap. 12.

There is a set of possible combinations, (x_1, x_2), that defines the set of feasible bids, denoted by S. A typical restriction would be

$$x_1 + x_2 \leq 1$$

Nash listed the following requirements on the solution:

- *Pareto efficiency*: There should be no alternative solution that implies a strict improvement for one of the players without worsening the outcome for the other.
- *Symmetry*: If a problem is symmetric, the outcome should be symmetric.
- *Scale invariance*: The solution should be invariant with respect to changes of the zero point and the unit in which utilities are measured.
- *Contraction independence*: If the set of feasible solutions is contracted but the previously derived solution is in that set, this solution will remain.[36]

Nash proved that the unique solution that satisfies the above four conditions is the point (x_1^*, x_2^*) that maximises the product.

$$f(x_1, x_2) = u_1(x_1)\, u_2(x_2)$$

over the set of feasible points (x_1, x_2).

The decades that have passed since the publication of Nash's seminal work have seen a number of criticisms and ramifications.[37] Alternative solutions such as the egalitarian and the Kalai-Smorodinsky solution have been proposed. One way of approaching the problem of choosing between various solution concepts has been devised for two-party bargaining by van Damme.[38] The idea is to transform the discussion about which solution concept to apply into a meta-bargaining game with the aim of negotiating a procedure. Obviously, this meta-bargaining is also in need of some axiomatic specification that should be acceptable to both parties. Van Damme then shows that a very weak assumption of *risk sensitivity* is enough to make the meta-negotiation game converge. By this assumption it is understood that the payoff to a player will not decrease if the other player becomes more risk averse. Under these assumptions, it can be shown that the Nash solution will be an equilibrium of the meta-bargaining game, and that no higher value can be guaranteed to any of the players; in other words, the Nash solution is a maximin strategy for the meta-bargaining game. Further, if the players do not know in advance which game they are going to play, then the Nash solution is the optimum. This is obviously a strong argument for the Nash solution in situations characterised by a high degree of uncertainty, such as in a social contract situation.

---

[36] This is sometimes referred to as "independence of irrelevant alternatives".

[37] See Thomson (2010) for a survey and a collection of important publications in this development.

[38] van Damme (1986).

A different approach, based on evolutionary games, has been used by Young.[39] The conclusion from this analysis is that the resulting division is close to the Nash solution in homogeneous populations.

Turning now to the *strategic model*, it has appeared in different shapes over the years. In general, we imagine bargaining processes to be protracted, with alternating bids from the parties involved. An early reference for this line of thinking is work done by Zeuthen in the 1930's, further developed by Harsanyi,[40] and later by Ståhl and Rubinstein. The essential building block in this family of models is that the parties involved present alternating bids—a process that continues until they have reached a mutually satisfactory solution. What drives the process is that time is considered valuable; the sooner there is an agreement, the better it is for both parties. One way of describing this pressure is to assign a discount factor to the process, which makes the pie to be divided shrink over time.

This approach defines a sequential game between the parties involved, and it is natural to ask what the possible Nash equilibria of this game are. It turns out that, in the absence of additional assumptions, there is an infinite set of equilibria. If it is required, however, that an optimal strategy be optimal at every decision point of the decision tree—what is referred to as *subgame perfection*[41]—there is a unique optimal strategy, and the Nash equilibrium of the sequential game will under certain conditions coincide with Nash's solution of the axiomatically defined bargaining game.[42] There is no need for actual bargaining and sequential offers in this strategic description. Both parties are able to compute the outcome, and they will agree on this outcome without delay if they behave rationally.

On closer inspection, it turns out that the conclusion is not quite as robust as one would hope, particularly not in the case of more than two players.[43] A necessary caveat is also that integer effects can sometimes lead to unexpected outcomes. But for the strategic approach as for the axiomatic, the Nash solution remains a strong candidate.

The axiomatic approach and the strategic approach are different but of course related. The purpose for which the model will be used in the present context is descriptive, without reference to what can be considered desirable from an ethical point of view. In this perspective, explanatory power and economy of assumptions are important criteria of choice. The above survey justifies the Nash solution as the archetype for use in bargaining analysis.

---

[39] Young (1993).

[40] Zeuthen (1930), Harsanyi (1956). Ståhl refined such a model in the late 1960s, which was later discovered independently by Rubinstein (Ståhl 1972; Rubinstein 1982). For a comparison between the two approaches, see Ståhl (1994).

[41] Selten (1965).

[42] For various assumptions leading to this convergence of the two solutions, see Osborne and Rubinstein (1990), Chap. 4.

[43] See Thomson (2010), Sect. 11.

2.5   The Mathematics of Oscillations*

It was noted above that negative feedback is necessary but not sufficient for stability. The classical example of a system that shows the insufficiency of negative feedback is the pendulum, governed by the equation

$$\ddot{x} + \omega^2 x = 0$$

This equation describes a sinusoidal motion without damping. In practice, the motion will be dampened because energy is dissipated via friction, and the amplitude of the pendular movement will approach zero.

If energy is fed into the system, a persistent periodic motion can be maintained. In Sect. 1.1, three paradigmatic examples of periodic motion were presented. These three examples refer to situations in which, respectively,

- the system is forced from outside by a periodic motion,
- the system is influenced from outside by a constant force that interacts with the system to generate a periodic motion, or
- the system generates a periodic motion autonomously (no external force).
  These examples will now be explained in turn.

Periodic External Force
The example presented in this category refers to temperature variations between day and night, between seasons and over a longer time horizon, between average climate and ice ages. Conceptually, this is the simplest category; there is nothing surprising about periodic motion in a system that is periodically forced from outside. A simple example is a system with some inertia being forced by a periodic motion with angular frequency ω:

$$\dot{x} + ax = \sin(\omega t), x(0) = 0, a > 0$$

The solution of this equation is given by

$$x(t) = (a \sin \omega t - \omega \cos \omega t) / (a^2 + \omega^2) + \omega e^{-at} (a^2 + \omega^2)$$

The first term follows the outer motion with a time delay, whereas the second is a transient that will eventually die out.

As explained in Sect. 1.1, there are several periodic motions superimposed in the mechanism generating ice ages. When these work in the same direction, ice ages are triggered. The effect is reinforced by a positive feedback mechanism, namely the albedo effect from ice that increases the amount of energy reflected and contributes to further cooling.

Non-periodic External Force
The generation of periodic motion by a non-periodic external force is illustrated by a reed in a river. At low speeds, the flow past the reed is *laminar*, that is, the speed at a

given point in the water is always the same. *Periodic* motion appears as the speed of the water reaches a certain threshold value; in this case, the flow speed at a given point varies with time but in a regular mode. If the flow speed is increased further, the flow becomes *turbulent*, and the speed becomes unpredictable.

Hydrodynamic motion is governed by the Navier-Stokes equation, a nonlinear partial differential equation for the speed vector $\vec{u}$ and the pressure p. The critical parameter of the flow is the Reynolds number Re, which is defined as the product of the speed of the fluid ($|u|$, the absolute value of u) and a typical length, L,—in this case the dimension of the reed—divided by the viscosity of the fluid, μ:

$$Re = u \cdot L/\mu$$

For a given fluid and spatial configuration, L and μ are constant, and the speed of the flow, u, becomes the driving parameter. Detailed analyses of the Navier-Stokes equation have shown that periodic vortex shedding appears at the critical value of $Re = 46$.[44] The birth of periodic motion as a function of an external parameter in the differential equation is referred to as a *Hopf bifurcation*. For higher values of the Reynolds number, around 200, transition to full turbulence occurs.

Autonomous Oscillations

The examples that were used in Chap. 1 to illustrate the existence of autonomous oscillations—predator-prey populations and business cycles—share certain characteristics. In both cases, flows of energy or goods circulate between subpopulations. The response of one component to the actions of other components exhibits some inertia or time delay, which is the reason why oscillations appear.

In the case of *predator-prey interaction*, the original model suggested by Lotka and Volterra involved only two species—one predator and one prey. The assumptions behind the model are simple. In the absence of other species, each population has a natural rate of growth ($\alpha > 0$, $-\delta < 0$ for the prey and predator, respectively). The likelihood of an encounter between predator and prey and a successful catch is assumed to be proportional to the product of the populations, resulting in population growth changes $-\beta xy$ and γxy ($\beta > 0$, $\gamma > 0$) for prey and predator. The resulting system of differential equations is

$$\begin{cases} \dfrac{dx}{dt} = \alpha x - \beta xy \\[2mm] \dfrac{dy}{dt} = \gamma xy - \delta x \end{cases}$$

This system of equations gives rise to cyclical variations in the population sizes. The model suffers from structural instability in the sense that a change of initial conditions will in general lead to permanent changes in population sizes. It is also too simplistic; real-life food webs are most often much more complex. The Canadian

---

[44]For a sequence of increasingly detailed analyses, see Hammond and Redekopp (1997), Pier (2002), Barkley (2006) and Sipp and Lebedev (2007).

ecosystem of lynx and snowshoe hare has been studied in depth and found to comprise a number of predators besides the lynx, such as foxes, wolves, and birds of prey. The hare has competitors at the same level of the food web (squirrels, ptarmigans) and is also dependent on the productivity of the vegetation. In spite of this complexity, the population variations can be modelled with reasonable accuracy using two dimensions for the lynx and three for the hare. In other words, the hare is regulated both from below and above by the vegetation and a variety of predators, whereas the lynx is regulated only from below, and then primarily by the hare.[45]

As for *business cycles*, both aggregate output and its main components exhibit significant fluctuations, as do other variables, such as employment, productivity and stock prices. Different mechanisms have been suggested as internal, or endogenous, sources of these fluctuations—goods production, financial investments, or interaction between wage earners and capital owners. The basic mechanism is the same; there is some deviation from a hypothetical equilibrium, and one group reacts to this deviation by changing its behaviour—increasing production, assuming a more aggressive posture in negotiations or taking greater risks. In an ideal market, such actions are uncoordinated, and an overreaction may occur because of the time delays from decision to concrete results at the aggregate level. This leads to a deviation in the other direction, and so the cycle may be closed. Obviously, the speed of adjustment is a central parameter in this context.

Kaldor's original model (1940) was based on a few simple assumptions concerning consumption, propensity to save, and investment. The dynamical behaviour of the system resembles that of the predator-prey interaction. The model has been developed in various directions towards higher levels of sophistication. A standard result is that either one or three steady states exist, and that endogenously generated oscillations occur in the presence of only one unstable equilibrium. An analysis of the dynamic properties of the system when three steady states exist shows the existence of a limit cycle surrounding the three steady states—in summary a highly complex dynamic structure.[46]

It can be shown formally that the introduction of a pure time delay in a classical Kaldor model also leads to a Hopf bifurcation and consequently to periodic motion, corresponding to a business cycle.[47]

2.6 The Game of Marbles*

The game of marbles sketched in Chap. 1 gives rise to a random walk, sometimes called *the gambler's ruin*. There are two players, *1* and *2*, who stake one marble each in each round, and the winner gets both. The probability of winning in each round is

---

[45] Stenseth et al. (1997).

[46] Bischi et al. (2001).

[47] Kaddar and Talibi Alaoui (2008).

p_1 and $p_2 = (1 - p_1)$, respectively. The game is over when one of the players has run out of marbles.

This random walk gives rise to a Markov chain with two absorbing states, that is, when either of the players has no more marbles. The outcome depends on the number of marbles that the players own at the outset and their relative skills.[48]

Let n_1 and n_2 be the initial holdings and set $N = (n_1 + n_2)$. Further, denote the odds ratio p_1/p_2 by r. Then the following results can be derived:

- If the two players are equally skilful, the probability of winning is proportional to the number of marbles at the outset:

$$P(1 \ wins) = n_1/N$$

$$P(2 \ wins) = n_2/N$$

- If player 1 is more skilful than player 2 $(r > 1)$, the probabilities of 1 and 2 winning are given by the expressions

$$P(1 \ wins) = \left(r^N - r^{N-n_1}\right)/\left(r^N - 1\right)$$

$$P(2 \ wins) = \left(r^N - r^{N-n_2}\right)/\left(r^N - 1\right)$$

The first result is intuitively clear; the probability of winning is proportional to the number of marbles in the player's possession when the game starts. If player 1 has 5 marbles and player 2 has 50, player 1 will win approximately 9 games out of 100 and player 2, 91 games.

If 1 is the more skilful player, this higher skill may compensate for the disadvantage. The r value that equalises chances of winning can be obtained from the equation

$$\left(r^N - r^{N-n_1}\right)/\left(r^N - 1\right) = 1/2 \tag{2.1}$$

This yields approximately $r = \sqrt[5]{2}$, given the values assumed. Player 1 must consequently be around 15 per cent more skilful than 2 in order to compensate for the disadvantage in assets.

The third case highlighted in Chap. 1—a series of games in which both players start with 5 marbles and receive another 5 marbles before each new game starts—is easily analysed. If they are equally skilful, the winner of the first round will start the second one with 15 marbles against 5 and so stands a 75 per cent chance of also winning the second game. This imbalance is reinforced with each new round played.

The stochastic nature of this game makes it possible for the disadvantaged player to win; all statements on outcomes refer to expected values. In the final variant, for

---

[48] For the full analysis, see Kemeny and Snell (1976), Sect. 7.1.

instance, there may be a swap of positions at any time during the sequence, but the likelihood of this occurring diminishes with every round played.

In summary, this simple example illustrates the three main categories of the origins of inequality:

- Situations in which inequality is generated by environmental factors.
- Situations where both the characteristics of the players and the game itself as well as environmental factors contribute to inequality. The threshold value at which these two groups of factors balance each other is given by expression (2.1).
- Situations in which inequality is internally generated by the rules of the game or the characteristics of the players, or a combination of these.

Bibliography

Atkinson, A. B. (1983). *The economics of inequality* (2nd ed.). Clarendon Press.

Barkley, D. (2006). Linear analysis of the cylinder wake mean flow. *Europhysics Letters, 75*, 750–756.

Bischi, G. I., et al. (2001). Multiple attractors and global bifurcations in a Kaldor-type business cycle model. *Journal of Evolutionary Economics, 11*, 527–554.

Björklund, A. (1993). A comparison between actual distributions of annual and lifetime income: Sweden 1951–89. *Review of Income and Wealth, 39*(4), 377–386.

Burkhauser, R. V., & Poupore, J. G. (1997). A cross-national comparison of permanent inequality in the United States and Germany. *The Review of Economics and Statistics, 79*(1), 10–17.

Cowell, F. A. (2000). Measurement of inequality. In A. B. Atkinson & F. Bourguignon (Eds.), *Handbook of income distribution* (Vol. 1). North-Holland.

Dawkins, R. (1976). *The selfish gene*. Oxford University Press.

Dobzhansky, T. (1958). Species after Darwin. In S. A. Barnett (Ed.), *A century of Darwin* (pp. 19–55). Heinemann.

Dover, G. (1982). Molecular drive: A cohesive mode of species evolution. *Nature, 299*, 111–117.

Durlauf, S. N., & Young, H. P. (Eds.). (2001). *Social dynamics*. MIT Press.

Easley, D., & Kleinberg, J. (2010). *Networks, crowds and markets*. Cambridge University Press.

Emery, F. E. (Ed.). (1969). *Systems thinking*. Penguin Books.

Foster, D., & Young, H. P. (1990). Stochastic evolutionary game dynamics. *Theoretical Population Biology, 38*(2), 219–232.

Fudenberg, D., & Tirole, J. (1991). *Game theory*. MIT Press.

Gould, S. J., & Lewontin, R. C. (1979). The spandrels of San Marco and the Panglossian paradigm: A critique of the adaptationist programme. *Proceedings of the Royal Society B, 205*(1161), 581–598.

Hammond, D. A., & Redekopp, L. G. (1997). Global dynamics of symmetric and asymmetric wakes. *Journal of Fluid Mechanics, 331*, 231–260.

Harsanyi, J. C. (1956). Approaches to the bargaining problem before and after the theory of games: A critical discussion of Zeuthen's, Hicks', and Nash's theories. *Econometrica, 24*(2), 144–157.

Hirshleifer, J. (1977). Economics from a biological viewpoint. *The Journal of Law and Economics, 20*(1), 1–52.

Hofbauer, J., & Sandholm, W. H. (2011). Survival of dominated strategies under evolutionary dynamics. *Theoretical Economics, 6*, 341–377.

Jablonka, E., & Lamb, M. J. (2020). *Inheritance systems and the extended synthesis*. Cambridge University Press.

Jenkins, S. P., & Van Kerm, P. (2009). The measurement of economic inequality. In B. Nolan et al. (Eds.), *The Oxford handbook of economic inequality.* Oxford University Press.

Kaddar, A., & Talibi Alaoui, H. (2008). Hopf bifurcation analysis in a delayed Kaldor-Kalecki model of business cycle. *Nonlinear Analysis: Modelling and Control, 13*(4), 439–449.

Kaldor, N. (1940). A model of the trade cycle. *Economic Journal, 50,* 78–92.

Kemeny, J. G., & Snell, J. L. (1976). *Finite Markov Chains* (2nd ed.). Springer.

Kuhn, H. W., & Nasar, S. (2002). *The essential John Nash.* Princeton University Press.

Logemann, H., & Ryan, E. P. (2014). *Ordinary differential equations. Analysis, qualitative theory and control.* Springer.

Mantel, R. (1974). On the characterization of aggregate excess demand. *Journal of Economic Theory, 7,* 348–353.

Maynard Smith, J. (1982). *Evolution and the theory of games.* Cambridge University Press.

Maynard Smith, J., & Price, G. R. (1973). The logic of animal conflict. *Nature, 246,* 15–18.

McNamara, J. M., & Weissing, F. J. (2010). Evolutionary game theory. In T. Székely et al. (Eds.), *Social behaviour: Genes, ecology and evolution.* Cambridge University Press.

Milo, D. S. (2019). *Good enough. The tolerance for mediocrity in nature and society.* Harvard University Press.

Nash, J. (1950a). *Noncooperative games.* Ph.D. dissertation, Department of Mathematics, Princeton University. Reprinted as chapter 6 of H.W. Kuhn, & S. Nasar (2002), The Essential John Nash. Princeton: Princeton University Press.

Nash, J. (1950b). The bargaining problem. *Econometrica, 18*(2), 155–162. Reprinted as chapter 4 of H.W. Kuhn & S. Nasar (2002), The Essential John Nash. Princeton: Princeton University Press.

Nash, J. (1953). Two-person cooperative games. *Econometrica, 21*(1), 128–140. Reprinted as chapter 8 of H. W. Kuhn, & S. Nasar (2002). The Essential John Nash. Princeton: Princeton University Press.

Osborne, M. J., & Rubinstein, A. (1990). *Bargaining and markets.* Academic.

Pier, B. (2002). On the frequency selection of finite-amplitude vortex shedding in the cylinder wake. *Journal of Fluid Mechanics, 458,* 407–417.

Rapoport, A., & Guyer, M. (1966). A taxonomy of 2x2 games. *General Systems, 11,* 203–214.

Ravallion, M. (2012). Troubling tradeoffs in the Human Development Index. *Journal of Development Economics, 99,* 201–209.

Rubinstein, A. (1982). Perfect equilibrium in a bargaining model. *Econometrica, 50,* 97–109.

Sandholm, W. H. (2010). *Population games and evolutionary dynamics.* MIT Press.

Scarf, H. (1960). Some examples of the global instability of the competitive equilibrium. *International Economic Review, 1,* 157–172.

Selten, R. (1965). Spieltheoretische Behandlung eines Oligopolmodells mit Nachfrageträgheit. *Zeitschrift für die gesamte Staatswissenschaft, 121,* 301–324.

Shorrocks, A. F. (1978). Income inequality and income mobility. *Journal of Economic Theory, 19,* 376–393.

Sipp, D., & Lebedev, A. (2007). Global stability of base and mean flows: A general approach and its applications to cylinder and open cavity flows. *Journal of Fluid Mechanics, 593,* 333–358.

Skyrms, B. (2001). The stag hunt. *Proceedings and Addresses of the American Philosophical Association, 75*(2), 31–41.

Sonnenschein, H. (1972). Market excess demand functions. *Econometrica, 40,* 549–563.

Ståhl, I. (1972). *Bargaining theory.* EFI, Stockholm School of Economics.

Ståhl, I. (1994). The Rubinstein and Ståhl bargaining models. A comparison and an attempt at a synthesis. *EFI research paper* 6535, Stockholm School of Economics.

Stenseth, N. C., et al. (1997). Population regulation in snowshoe hare and Canadian lynx: Asymmetric food web configurations between hare and lynx. *Proceedings of the National Academy of Sciences USA, Ecology, 94,* 5147–5152.

Taylor, P., & Jonker, L. (1978). Evolutionarily stable strategies and game dynamics. *Journal of Applied Probability, 16,* 76–83.

Tegmark, M. (2014). *Our mathematical universe*. Alfred A Knopf.

Thomson, W. (Ed.). (2010). *Bargaining and the theory of cooperative games: John Nash and beyond*. Edward Elgar.

van Damme, E. (1986). The Nash bargaining solution is optimal. *Journal of Economic Theory, 38*(1), 78–100.

Weibull, J. W. (1995). *Evolutionary game theory*. MIT Press.

Williams, G. C. (1966). *Adaptation and natural selection: A critique of some current evolutionary thought*. Princeton University Press.

Wynne-Edwards, V. C. (1962). *Animal dispersion in relation to social behaviour*. Oliver and Boyd.

Young, H. P. (1993). An evolutionary model of bargaining. *Journal of Economic Theory, 59*(1), 145–168.

Young, H. P. (1998). *Individual strategy and social structure. An evolutionary theory of institutions*. Princeton University Press.

Zeuthen, F. (1930). *Problems of monopoly and economic warfare*. Routledge and Kegan Paul.

Chapter 3
Life-Cycle Development

At any instant during the life cycle, an individual carries a range of possibilities for the future. How this spectrum of possibilities changes over time depends both on the actual state—the "momentum"—and the environment. The general direction of movement during most of the life cycle goes from total dependence on the environment—the pregnant mother and the close family circle—towards increasing autonomy.

External factors may be positive stimuli or negative disturbances of various kinds. Such external influences are not random—their effects may vary systematically with the current state. Individuals below average may run greater risks of disturbances, the damage may be greater if risks materialise, and their capacity to cope may be more limited. As a consequence of such feedback mechanisms, inequalities may accumulate over the life cycle.

Outcomes can be measured in a number of dimensions—education, health, occupation, social status, earnings, or wealth. There is a tendency for such outcomes to co-vary,[1] but correspondence is not one-to-one. Which measure of inequality is used may be important to what conclusions are drawn.

In the public debate on inequality, a distinction is often made between opportunities and outcomes. As will be illustrated in the present chapter, this distinction is difficult to maintain as the individual passes through the successive stages of a life cycle. The outcome of one stage sets the initial conditions of the next, which to a considerable extent defines the limits of the opportunities available. In this sense, the individual constantly carries a memory of his or her earlier life, for which the metaphor of *capital* is appropriate.[2]

---

[1] See, for example, Hanushek et al. (2015) on the relation between income and skills and Oreopoulos and Salvanes (2011) on the relation between education and non-pecuniary outcomes.

[2] This concept forms the basis of Chap. 4.

3.1   Inheritance as a Source of Inequality

From its total dependence on the mother during the foetal phase and immediately after birth, the child is exposed to an ever-widening circle of influences while growing up. Environmental factors differ in character and strength, and their impact depends also on the current state of development of the individual. The interaction between the subject and the environment is therefore highly complex, and simple statements on the relative importance of various factors are liable to mislead.

Intergenerational Transmission
A child inherits not only genes from its parents but also a host of other factors of influence—siblings and other relatives, housing and the close neighbourhood, schools and others. This bundle of genes and environmental factors explains the *intergenerational transmission* that is an important source of inequality, its persistence and growth. High mobility in a society implies that these transfers between generations are of limited importance. The strength of these transfers—their *persistence*—can be defined as a number between 0 and 1 indicating how much of an increase in parental talent, education, income, etcetera, is forwarded to the children.[3]

Several forms of resources may be transmitted. Different labels are used. One tripartite categorisation is *material*, *relational* and *embodied capital*, corresponding to *physical*, *social* and *human capital* in economic terminology. Obviously, the mechanisms and conditions of transmission differ between and within these categories.

Even when limited to one of these forms, for instance income, several definitions of mobility or persistence are possible. Jäntti and Jenkins discern four main alternatives[4]:

- The definition of mobility may refer to a person's *position* in the ranking of income (referring to two points in time for the same person or parents and children).
- The definition may refer to *changes in income* in absolute terms (same person or parents versus children).
- *Average income over a longer period* is less volatile than annual income. The difference between these two can serve as a measure of mobility.
- Alternatively, the same difference can be used as a measure of *income risk*. The construction is similar, but the normative implications are different.

The choice of measure is important for the outcome, for instance when comparing countries. The ranking with respect to mobility will in general depend on the perspective and measure chosen.

---

[3]For an overview of transmission theories, see Piketty (2000); specifically on premodern societies, see Borgerhoff Mulder et al. (2009) and Bowles et al. (2010). Jäntti and Jenkins (2015) review both intergenerational and intragenerational mobility. Piketty and Zucman (2015) is devoted to the inheritance of wealth.

[4]Jäntti and Jenkins (2015).

With the necessary caveats, results from a number of comparative studies of intergenerational transfers nonetheless provide an overall picture of their importance in different countries and how it has changed over time. Some important stylised facts are[5]:

- In *premodern societies*, intergenerational transmission is highest where material assets dominate, that is, in pastoral and agricultural societies. Reported transmission factors are around 0.4, as compared to 0.2 in hunter-gatherer and horticultural societies.[6]
- In modern societies, intergenerational persistence in *education* is generally higher in low-income than in high-income countries, with middle-income countries in between. The average rate of persistence in the low-income countries of South Asia and sub-Saharan Africa is 0.57 and 0.54, respectively, implying that an additional year of parental education is on average associated with over half a year of additional education for their children. The corresponding global average for middle-income countries is just over 0.4, and for high-income countries, 0.32.[7] A detailed study of educational mobility in Africa illustrates the complexity of causal relations in mobility patterns.[8] Geographic variability is high, including at the regional level, affected by proximity to urban centres and to the coast. The colonial heritage plays a role via investments in infrastructure, as do Protestant mission activities. But individual history is also important; childhood exposure to high-mobility areas increases mobility.
- Intergenerational persistence in *incomes* exhibits a similar pattern—higher in low-income countries and vice versa. There is substantial variation within these categories, however. Mobility is high in Australia, Canada and the Nordic countries but substantially lower in the United Kingdom, the United States and southern Europe. It is also higher in Asia than in Latin America and Africa.[9]
- Intergenerational persistence of *income in high income strata* and in *wealth* is generally strong, irrespective of the general distribution of income.[10] The inherited part of private wealth is currently above 50 per cent in major European countries.[11]
- Intergenerational persistence in *health* has been investigated using birthweight or self-reported health as indicators.[12] Data from the United States indicate that persistence is significantly lower than persistence in income. There is strong genetic causation for certain diseases, but they are rare.

---

[5]Measurements of social class mobility are fraught with methodological problems and are omitted here. For more detailed information on a choice of OECD countries, see Ermisch et al. (2012).

[6]Borgerhoff Mulder et al. (2009).

[7]Nybom (2018).

[8]Alesina et al. (2021).

[9]Nybom (2018).

[10]De Nardi (2004), Björklund et al. (2012).

[11]Piketty and Zucman (2015).

[12]Currie and Moretti (2007) (birthweight), Halliday et al. (2018) (self-reported health).

- During the twentieth century, *educational* mobility increased in most countries following the expansion of mandatory basic education. *Income* mobility increased in the Nordic countries, decreased somewhat in the United Kingdom and the United States, and was stable or insignificant elsewhere. Rising inequalities since the 1980s have so far not left any visible imprints on mobility.[13]
- Higher inequality is associated with lower intergenerational mobility in both education and incomes.[14] This negative correlation between inequality and mobility, nicknamed *the Great Gatsby curve*, can also be observed across regions within countries.[15]

By way of illustration, consider Fig. 3.1, showing the relationship between income inequality in childhood, and the intergenerational persistence in income between parents and children (when children have reached adult age).

These general results are important as a point of departure for both further analysis and for policy but say nothing about the mechanisms involved. The main categories are genetic factors (see the following section) and environment. The strength of purely genetic factors can be assumed to be reasonably constant over

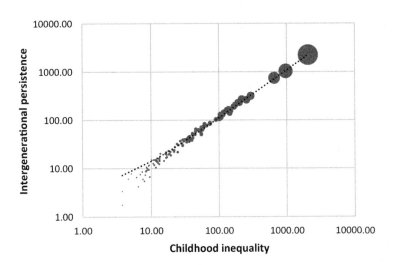

Fig. 3.1 "The Great Gatsby curve" in Sweden. Relationship between childhood inequality (Gini coefficient) and intergenerational persistence in labour market regions in Sweden. Data points are weighted by population size in the regression. The straight line represents "the Great Gatsby curve". Source: Brandén (2019)

---

[13] Nybom (2018).

[14] On education, see Narayan et al. (2018); on incomes, an early contribution to the literature is the comparison between the U.S. and Sweden by Björklund and Jäntti (1997). See further Björklund and Jäntti (2009), Corak (2013) and Blanden (2009).

[15] Chetty et al. (2014) (U.S.), Güell et al. (2018) (Italy), Fan et al. (2018) (China), Corak (2017) (Canada), and Heidrich (2017) and Brandén (2019) (Sweden).

time and space, so differences in these dimensions can be ascribed to differences in the environment.

As for environmental factors, several studies indicate that the system of education is an important link between the general level of inequality and both education and income mobility.[16] Inequality may also affect mobility via segregation, as higher inequality tends to increase residential segregation.[17]

In regard to wealth, detailed studies indicate that gifts and bequests are the dominant channel of transmission. The genetic component is weak, particularly in the highest strata.[18] A detailed study of adopted children in Sweden shows that prior to any inheritance, there is a substantial role for environmental but a limited role for pre-birth factors. Adoptive parental wealth is important via bequests, whereas the role for child earnings, education, and savings rates is weak. In fact, savings rates are lower among children of wealthier parents.[19]

Genetics and Epigenetics

The longstanding debate over nature versus nurture has always been heated, often driven by ideology more than by facts. In conservative circles, the tendency has been to ascribe a large part of personality to innate entities, originally blood, later genes. At the other extreme, Lysenkoism—the official ideology during the Soviet era— claimed that personality can be more or less completely modelled under the influence of judiciously chosen environmental inputs. The prevailing scientific consensus sees human development as the result of a complex interplay between genes and the environment, and it is the environment that in the final instance determines the relative importance of these two categories.

A common phraseology in research publications on the issue of genes versus environment in explaining a given trait is that "genes explain X per cent of the variation and the environment *(100 − X)* per cent". Such a statement cannot be generalised and will most likely be misunderstood. What it means is that *under the given environmental conditions*, total variation can be split according to the figures presented. If a population of children grows up under harsh conditions, with little or no emotional or intellectual stimuli, total variation will be small, and the genetic component will dominate whatever variation there is. In a more beneficial environment, total variation will be larger, and the relative importance of environmental factors will increase.

A common approach to the question of the importance of inheritance has been to study siblings, twins and adopted children—categories that represent different combinations of genes and family environment. All these methodologies suffer

---

[16] Narayan et al. (2018) show that mobility is higher in countries with a relatively higher public budget for education. See also Brunori et al. (2013).

[17] Chetty et al. (2014), Durlauf and Seshadri (2017).

[18] Björklund et al. (2012), Adermon et al. (2018), Collado et al. (2020).

[19] Black et al. (2020).

from a number of problems, however.[20] So-called "identical" twins are not identical, differing for instance in birth-weight. If twins follow different life courses because of some external factor, for instance the death of their parents, one cannot exclude the possibility that such an event also had deeper consequences. Being a twin or being adopted is a rather special life condition, which raises a question about the generalisability of results obtained.

The mapping of the human genome has opened up new possibilities for analysing the link between genes and traits or behaviour. The method is based on the identification of a number of DNA sequences that could be coupled to a certain trait, so-called candidate genes, whereupon a large number of correlations are computed.[21]

There are also measurement problems with some of the central variables in focus. Cognitive skills are estimated in aptitude tests normally carried out when the child reaches school age or even later, but then the effect of the environment has already left an imprint on the child's personality and skills. The effects of the early environment, from conception till the age of five, on cognitive skills measured in the late teens is of the same order of magnitude as the effect of the educational system, according to current estimates.[22] An inherent problem with the DNA-based approach is that complex outcomes such as cognitive or non-cognitive skills can be expected to involve a large number of genes.

The long-standing debate on genes versus environment has undergone radical changes in recent decades, following substantial progress in research on *epigenetic processes*.[23] Genes and environment interact through a large part of the life cycle, which makes traditional statements on what percentage of total variation should be ascribed to genes or environment less meaningful. Factors that are known to affect genetic expression include, for instance, food, chemical substances in the environment, drugs, physical exercise, educational methods, general stress during pregnancy and early childhood, and work complexity.[24] Studies of both humans and primates (rhesus macaques) indicate epigenetic effects of social status.[25]

Some of the changes induced are reversible, whereas others have long-lasting, even lifelong effects on individual capabilities. There is evidence of epigenetic inheritance, that is, when phenotypic variations that do not stem from variations in DNA base sequences are transmitted to subsequent generations of cells or organisms.[26] Epigenetic mechanisms can represent an adaptive advantage, but it is difficult to say anything in general about their effect on inequality.

---

[20] See, for example, Sandewall et al. (2014) on twin research.

[21] For an example, see Lee et al. (2018), who claim that 7–10 per cent of cognitive skills are explained by genetic variation.

[22] Almond and Currie (2011).

[23] Moore (2015) offers an overview of the field.

[24] Ibid., Chap. 8; on work complexity, see Fratiglioni et al. (2020).

[25] Cole et al. (2007, 2012).

[26] Jablonka and Raz (2009), Jablonka and Lamb (2020).

Research on epigenetic processes is a rapidly expanding field, and it is too early to tell to what extent it will change our views on individual development. Nonetheless, it is already clear that the room for stimulating it is greater than what most of the twentieth-century genetic research has conveyed.

3.2   Prenatal and Early Childhood Development

In a secular perspective, improvements of human health have been the result of a handful of large-scale changes. Increased agricultural productivity and general economic growth during the eighteenth and nineteenth centuries permitted the elimination of chronic malnutrition, which had till then threatened health and limited population.[27] These benefits of the industrial transition have to be balanced against the health hazards associated with urbanisation, however. Cities suffered from recurrent outbreaks of infectious diseases, and it was not until the general expansion of water sanitation and sewerage systems around 1900 that city populations grew steadily without an inflow from the countryside.[28] On the other hand, urbanisation was a protracted process, and the urban fraction of the total population remained limited for a long time, so high mortality in the cities had a limited effect on overall longevity.

During the twentieth century, other prophylactic measures such as vaccination programmes and health screening contributed to improved health and increased longevity. In general, improvements were more important in the lower socioeconomic strata. Antibiotics and improved health care have been important in the post-World War II era, but in an environment where mortality had already been substantially reduced by other measures.

These changes have led to a general increase in longevity. In the industrialised countries, development has flattened out in recent decades, leading to general convergence in longevity. But significant differences remain within all countries, and there is no clear movement in the direction of increased equality in this respect. Within the European Union, there is diversity between the member countries, and in the United States inequality in longevity has been rising.[29,30]

---

[27] McKeown (1976) is the classical reference for this view. Later contributions include Fogel et al. (1983) and Fogel (1986). Preston (1975) warns against exaggerating the role played by economic development. See also Bengtsson and van Poppel (2011).

[28] On the demographic effect of these infrastructural changes, see Cutler and Miller (2005) (U.S.), Alsan and Goldin (2015) (U.S.), Kesztenbaum and Rosenthal (2017) (France), Gallardo-Albarrán (2020) (Germany) and Helgertz and Önnerfors (2019) (Sweden).

[29] Debón Aucejo et al. (2017) (EU), Bosworth et al. (2016) (U.S.).

[30] A general discussion of various explanations for decreases in mortality can be found in Cutler et al. (2006).

The Prenatal Phase

The importance of perinatal and early childhood environment for individual development has been generally recognised in recent decades.[31] In fact, inequality starts to develop even before conception. The medical profession has become increasingly aware of the importance of preconception health to birth outcomes. Factors such as diet and nutrition, smoking, alcohol and obesity may have consequences even before conception.[32]

Turning to development in the womb, inequality is affected not only by genetic differences but also by environmental stress factors of various origins—psychic stress in the pregnant mother, use of alcohol and tobacco, fasting, noise and toxic substances in the environment, among others. Such factors may vary systematically with the socioeconomic background of the mother and thus contribute to the maintenance or increase of inequality across generations. They also affect capabilities for learning later in life and therefore contribute to the persistence and widening of gaps among adults.

A number of stress factors and disturbances may have lasting impacts on the development of the foetus and later in life. Some of the mechanisms involve hormonal systems and cell function, and some are epigenetic.[33] Indicators used in econometric studies such as cognitive skills or life-time income turn out to be more sensitive than traditionally used, purely medical ones, such as birth-weight. A common observation is that certain periods during foetal development are particularly sensitive to disturbances.

A general reservation concerning epidemiological studies of environmental effects on health is that there is a risk of publication bias; a study that identifies an effect is more likely to be published than a study with a negative result. On the other hand, this risk is gradually reduced as knowledge about the physiological mechanisms involved increases.

High stress levels in a pregnant woman may have negative consequences for the child's future health, cognitive skills and school performance.[34] Children of mothers with high and variable cortisol rates, which are particularly common in women with low incomes and a short education, are more sensitive to such stress factors.

Stress factors may be purely external or intervene more directly in the development of the foetus. Persons who were in the foetal stage during the Korean War suffered an impairment of both health and cognitive capacities, with consequences for educational performance, position in the labour market and other welfare indicators.[35]

The Chernobyl nuclear disaster in April 1986 caused a spike in *radiation* levels in Sweden, which were considered safe by radiation experts at the time. In a comprehensive study of Swedes born between 1983 and 1988, Almond et al. found that the cohort in utero during the Chernobyl accident had worse school outcomes than

---

[31] For overviews, see Almond and Currie (2011), Almond et al. (2018).

[32] Stephenson et al. (2018).

[33] Hoffman et al. (2017), Arima and Fukuoka (2019); on epigenetics, see Petronis (2010).

[34] Graignic-Philippe et al. (2014), Aizer et al. (2016).

[35] Lee (2014).

adjacent birth cohorts, and that this deterioration was largest for those exposed approximately 8–25 weeks post conception.[36] There was a clear dose-response pattern; damage among students born in regions that received more fallout was larger. Cognitive ability is consequently a more sensitive indicator of radiation damages than the clinical indicators previously used.[37]

Prenatal exposure to *chemical pollutants* such as nitrogen dioxide or tobacco smoke may have lasting effects on health and educational outcomes.[38] Phthalates, chemical compounds that are commonly used in packaging, hygienic articles, toys and floors, may leak into the environment and cause disturbances in sexual development or croup in a child if the mother is exposed during pregnancy.[39]

The pandemic called *the Spanish flu*, which hit the world in 1918, caused long-term health problems for those who were at the foetal stage, including impaired sight and hearing, and diabetes.[40]

Family ruptures that hit a pregnant mother affects mental health in the child.[41] Prenatal exposure to the death of a maternal relative increases the use of ADHD drugs during childhood and anti-anxiety and depression medications in adulthood. Family ruptures during pregnancy also depress birth outcomes and raise the risk of perinatal complications. Low-income mothers suffer greater stress exposure.

Malnutrition, forced or voluntary, is another source of developmental problems. This form of stress is common during wars and warlike conditions. Conditions in the Netherlands during World War II have been the object of a number of studies.[42] Persons who were in the foetal stage during the Dutch famine have displayed elevated risks for a number of health problems, such as obesity, type-2 diabetes and schizophrenia. An epigenetic link has been verified.[43]

In the Muslim world, the Koran requires that believers abstain from food and drink between sunrise and sunset during the month of fasting, *Ramadan*. The Koran admits of certain exceptions but on the other hand threatens with strong penalties, if this possibility is abused. This creates a potential risk for foetal development, which has been the object of numerous studies both in Muslim countries and among Muslim minorities elsewhere.[44] The most common indicators used are birth-weight

---

[36] Almond et al. (2009). Students from the eight most affected municipalities were 3.6 percentage points less likely to qualify for high school as a result of the fallout.

[37] A possible objection would be that the damage was caused more by the psychic reaction of the pregnant women than by radiation itself, but similar results have been obtained from Norwegian data on the effects of Soviet nuclear weapons tests during the 1950s and 1960s, where the distribution of the fallout in time and space was unknown to the population; see Black et al. (2019).

[38] Sanders (2012), Vieira (2015).

[39] Bornehag et al. (2015), Shu et al. (2018).

[40] Almond and Mazumder (2005).

[41] Persson and Rossin-Slater (2018).

[42] See Lumey et al. (2011) for a survey.

[43] Tobi et al. (2018).

[44] See Majid (2015) on Muslim countries; see further Almond et al. (2014) on the UK, Oosterbeek and van der Klaauw (2013) and Savitri et al. (2014) on the Netherlands, and Jürges (2015) on Germany (birth-weight only).

and cognitive skills. Because of the room for manoeuvre in interpreting the Koran, outcomes may depend on the attitudes of local religious leaders. In general, effects are stronger in Muslim countries. In Europe, there are weak or no effects on birth-weight but significant effects on cognitive capacities.

Chemical substances such as *alcohol, tobacco* and *narcotic drugs* imply obvious risks for the foetus when taken up by the pregnant mother. Alcohol abuse is associated with a broad variety of effects, so-called *foetal alcohol spectrum disorders* (FASDs).[45] Registered effects include a vast array of problems—growth inhibition, impaired cognitive capacity, cardiovascular diseases and metabolic disorders.[46] There also appears to be a negative effect on the ability to wait, an important determinant of success later in life.[47] The prevalence of FASDs in the United States has been estimated at 5 per cent of the population, and in Sweden at between 1 and 3 per cent.[48]

A large-scale experiment with increased alcohol availability was carried out in Sweden in 1967–68, when strong beer[49] was made available outside the state-managed alcohol monopoly in two regions, the rest of the country serving as control group. The labour productivity of those exposed in utero was followed up at the age of 32.[50] Compared to the surrounding cohorts, the exposed children had substantially lower cognitive and non-cognitive ability, worse educational outcomes (3.9 per cent lower probability of a high school exam) and worse labour market outcomes (24 per cent lower earnings). Effects on earnings were largest below the median. Males were more affected than females.

Tobacco and *narcotic drugs* have not been studied as intensively as alcohol, but there is evidence of similar damages to foetal development—impaired cognitive skills, deformities and elevated risk of disease in adult life. Epigenetic effects have been verified.[51]

Even relatively weak *adverse economic experiences* around birth may have long-lasting consequences in adult life. Being born during a recession implies an elevated risk of cardiovascular disease, increased sensitivity to adverse events at advanced ages, and increased mortality during the life cycle.[52] Effects are limited in size, but so are the disturbances.

---

[45] See Sokol et al. (2003) for a survey.

[46] Alati et al. (2006).

[47] Williams et al. (1994).

[48] May et al. (2018) for the U.S., Andreasson et al. (2020) for Sweden.

[49] Containing more than 3.5 per cent alcohol.

[50] Nilsson (2017).

[51] On tobacco, see Cnattingius (2004), Einarson and Riordan (2009); on drugs, Behnke and Smith (2013), Huizink (2014), Vassoler et al. (2014).

[52] Cardiovascular disease: Alessie et al. (2019); adverse events: Scholte et al. (2017); mortality: van den Berg et al. (2006), van den Berg et al. (2017).

Early Childhood

Early life years are as important to individual development throughout the lifecycle as the foetal phase, although the mechanisms involved are of course different.[53] Late health effects of early life adversity have been registered among nonhuman primates.[54] After birth, a child's development is still highly dependent on the mother but now also on the close environment. Parents from different socioeconomic groups tend to use different *styles in rearing* their children.[55] Parents with a higher status, in particular a higher education, tend to accord greater freedom of action to their children and to be less restrictive and punishing when socialising. The immediate environment also tends to be more stimulating and geared towards interaction and training of capabilities in such families. Health effects of the mother's level of education have been recorded.[56]

A stable observation is that mothers with a low socioeconomic status on average are *younger*.[57] Besides the educational gap, there is consequently a difference in general life experience.[58]

The *educational climate* is even more important to language learning.[59] Parents with a higher socioeconomic status talk more with their children, use a richer vocabulary and convey more information, for instance by explaining causal relationships.[60]

Stimulating environments increase the plasticity of the brain and leave visible traces in brain structures, not only among humans but also in other species.[61] This may reduce vulnerability to stress-related damages, aging and neurodegenerative diseases. Conversely, child poverty on average impairs working memory in young adults.[62]

Likewise, *stressful conditions* or events leave permanent traces in the brain. Epigenetic processes have been verified,[63] but other processes are also involved. Stress in the main carer, most often the mother, is liable to generate long-term consequences for the child. Poverty increases the risk of *maternal depression*, which in turn can weaken attachment and impair child development. This increases

---

[53] To the previously cited surveys, Almond and Currie (2011), Almond et al. (2018), can be added Currie and Rossin-Slater (2015).

[54] Conti et al. (2012).

[55] For international surveys, see Hoff et al. (2002), Bradley and Corwyn (2002), Lareau (2002).

[56] Chen and Li (2009).

[57] Nettle et al. (2011) (UK), McCall et al. (2015), Vikat et al. (2002) (teenage motherhood, Scotland and Finland, respectively).

[58] The direct causal effect of age seems to be limited, however; see López Turley (2003).

[59] Hoff (2006), Weisleder and Fernald (2013).

[60] Hart and Risley (2003) coined the term *The 30 Million Word Gap by Age 3*.

[61] Mohammed et al. (2002).

[62] Evans and Schamberg (2009).

[63] Harms et al. (2017).

the risk of future poverty.[64] There is evidence from all parts of the world to suggest a causal relationship between *physical abuse* in childhood and a range of mental disorders in adult life, such as drug use, suicide attempts, sexually transmitted infections, and risky sexual behaviour.[65]

Turning to the external environment,[66] it has been shown that *housing and neighbourhood* conditions can have lasting effects on a range of outcomes—health, educational level, teenage motherhood and earnings. Health effects have been shown both for mortality in general and for specific diseases.[67] Effects of housing are not strong but specific, beyond adverse socioeconomic conditions in general. The effect of neighbourhoods is cumulative.[68]

Even low levels of *pollution* can affect infants' health.[69] Socioeconomic status is important; children from poor or minority families are more likely to be affected.[70] Detailed studies have been carried out on particulate matter and various toxic substances, such as lead.[71] Exposure to lead in the air affects long-run outcomes, such as noncognitive skills, which in turn explain a sizeable share of the impact on crime and human capital formation. As in many other contexts, boys are more affected than girls.

Early child development, like the foetal phase, is not uniform but passes through sensitive or critical periods, and cognitive and non-cognitive capabilities follow different paths.[72] In general terms, early periods are particularly important. Nonetheless, it should be recognised that these windows can to some extent be reopened later in life and that the human brain remains malleable in adulthood.[73]

3.3   Preschool and School Years

It is clear from the previous sections of this chapter that as a child enters the public system of care and education—preschools and schools—it already carries a stock of experiences and imprints from its early life, positive as well as negative. Some are reversible, others indelible. Together with the genetic heritage, these external

---

[64]Reeves and Krause (2019).

[65]Norman et al. (2012).

[66]Graff Zivin and Neidell (2013) provides a survey of economic contributions to the field.

[67]See Dedman et al. (2001) for mortality in common diseases and Claussen et al. (2003) for cardiovascular disease risk.

[68]As well as the previously cited Chetty et al. (2014), see Galster et al. (2007) and Hedman et al. (2015).

[69]See the survey by Currie (2013).

[70]Currie (2009), Jans et al. (2018).

[71]Zhang et al. (2018), Grönqvist et al. (2020).

[72]Thomas and Johnson (2008).

[73]Hensch and Bilimoria (2012).

influences shape personality traits, health, and cognitive as well as noncognitive capacities. The general profile of the external factors acting on the child is not random but varies with socioeconomic background. About half of the inequality in the present value of lifetime earnings in the United States can be referred to factors determined by age 18,[74] and similar relationships can be found for other outcomes.

Education

Children's family environment is an important predictor of cognitive and socioemotional abilities. So is the public system of preschools and schools. School systems around the world represent different levels of ambition when it comes to equalising abilities in children. Even within the relatively homogeneous group of OECD countries, differences in literary proficiency between adults with highly and poorly educated parents vary widely.[75]

There is a tendency for ability gaps to persist or widen with age. How strong this tendency is in the absence of compensatory measures is difficult to ascertain, but even when such measures are present, gaps are liable to widen with age.[76]

Preschools are important. Participation in early childhood education of good quality has a lasting impact on indicators such as primary and secondary school progression, years of schooling, highest degree completed, employment, and earnings.[77] Universal preschool programmes tend to be more beneficial for children with an adverse socioeconomic background. There are no consistently different effects between boys and girls.

Whether the school system reduces gaps on entry or perpetuates, or even widens, them depends on both *institutional characteristics* and the *pedagogical regime*. Early academic tracking and school choice tend to widen the gaps.[78] It appears that the level—central or local—with which the main responsibility for the school system is vested is of some importance. Human capital is to a considerable extent a public good at the national level, and standard economic arguments indicate that decentralised decision-making on school budgets would lead to underproduction. France and Sweden are countries that have decentralised responsibility in this policy area in recent decades and that have also experienced a fall in educational performance.[79]

Quality is important in both preschools and schools. In statistical analyses, binary dummy variables and enrolment figures are often used because information on quality is difficult to obtain. This problem becomes particularly acute in countries with weak statistical institutions, as illustrated by a number of field studies. Drèze

---

[74] Heckman (2008).

[75] OECD (2017), Chap. 2. Differences are low in, for example, Japan, Korea and Estonia, and high in Israel, the UK and the U.S.

[76] See Cunha et al. (2006) and Cunha and Heckman (2009) for examples that refer to data on aptitude tests for 3- to 18-year-olds from the USA.

[77] Burger (2010), Dietrichson et al. (2020).

[78] Academic tracking: OECD (2017), Chap. 2; school choice: Musset (2012).

[79] As a statistical basis, this is obviously thin. For a more detailed discussion, see Hirsch (2016), Chap. 7.

and Sen report large distortions from India.[80] Although, on average, only 10 per cent of pupils aged 6–14 were registered as non-enrolled in the year 2004–5, the proportions of children able to perform the elementary operations of reading, writing and subtraction in the same year were no higher than 50, 64 and 43 per cent, respectively.[81] The gap is the combined result of misreporting and the low quality of education delivered. Chaudhury et al. studied surveys in which investigators made unannounced visits to primary schools and health clinics in six countries in Africa, Asia and Latin America. On average, about 19 percent of teachers and 35 percent of health workers were absent.[82] Many providers who were at their facilities were not working, so even these figures present too favourable a picture. In India for instance, a quarter of government primary school teachers were absent from school, but only half of the teachers present were actually teaching when investigators arrived at the schools.

Teaching methods matter. Within the pedagogical profession, a heated debate has been led over a number of decades on the pros and cons of different approaches. The main alternatives have been presented as a traditional, teacher-led regime and a pupil-oriented, or constructivist, approach. The main argument for the latter alternative has been that if the gap between what the pupils know and what the teacher attempts to convey is too large, no knowledge-building takes place. For this reason, knowledge should be constructed from below, based on the pupils' own active search. The problem with this argument is that the pupils are limited in their search by their current state of knowledge and they run the risk of getting trapped in an equilibrium of ignorance. Empirical studies have played a limited role in this debate, in spite of relatively clear results. The OECD PISA survey from 2015 indicates that the optimal approach is neither the traditional teacher-led regime nor the constructivist approach but what is appropriately called an *adaptive* approach.[83] This is still a regime that reserves a central role for the teacher but requires that teaching be strictly adapted to the pupils' current state of knowledge.

Of particular importance to the perspective of inequality is how different pedagogical approaches affect pupils with different backgrounds. There is clear evidence that the constructivist approach is unfavourable to children with a weak socioeconomic and educational background.[84]

Much of what has been said about early childhood development also holds for education, in the sense that an individual who is well equipped anywhere along the path of education stands a better chance of absorbing and profiting from an incoming flow of information. In the absence of conscious efforts to compensate for differences, the natural tendency will therefore be for differences to be reinforced. The positive conclusion is that investments in students with a weaker background will

---

[80]Drèze and Sen (2013).

[81]Desai et al. (2010).

[82]Chaudhury et al. (2006) (Bangladesh, Ecuador, India, Indonesia, Peru and Uganda).

[83]OECD (2016, 2018a).

[84]Chall (2000) and Mayer (2004) (both U.S.), Haeck et al. (2012) (Canada), Andersen and Andersen (2015) (Denmark).

raise not only their current level of knowledge but also the payoff to future investments.

At the lower end of the distribution, human capital accumulation may take more dramatic turns. A pupil who makes little or no progress may develop a negative attitude to learning itself, an anti-education regime, which of course impairs the process of learning further. This reaction, representing a critical turn or a *bifurcation point* in the process of human capital formation, is most common among boys.[85] It may also have a deleterious effect on health and more general social outcomes (see the following section).

Health

Obviously, ill health may have deleterious effects on a number of welfare-related outcomes—education, family formation, earnings and others. As noted earlier in this chapter, socioeconomic background is an important determinant of child health. Low socioeconomic status is associated with elevated risks of a variety of ill-health problems. Parents may intervene in response to health problems in the child, but the resources available and the knowledge about appropriate countermeasures will co-vary with socioeconomic status. There is ample empirical evidence to suggest that parental investments reinforce initial endowment differences.[86]

Gaps in health status, like cognitive abilities, tend to widen with age. This also holds when household income is kept constant, both for children and adults.[87]

There is a well-established link between education and health,[88] but opinions differ on the direction of causality and the mechanisms involved. Mental health problems in young people imply elevated risks of dropping out of school,[89] which in turn is a predictor of unemployment and criminal behaviour.[90] On the other hand, there is evidence of causal links from education to health in adult life.[91]

In order to get a more detailed picture of these causal links, it is necessary to identify the roles played by cognitive and non-cognitive skills, health endowments and family background.[92] Some of the health gaps observed in adult age that can be explained by education derive from selection into education based on early life endowments, while others are a causal effect of education. Personality traits are important. The selection component explains more than half of the observed difference by educational level in outcomes such as poor health, depression and obesity. On the other hand, education has an important causal effect on smoking rates and other health-related behaviours.

---

[85] Warrington et al. (2000), (UK data).

[86] Almond and Mazumder (2013).

[87] Case et al. (2002) (USA), Prus (2007) (Canada).

[88] Currie (2009), Cutler and Lleras-Muney (2010), Grossman (2008).

[89] Gustafsson et al. (2010).

[90] Bjerk (2012), Bäckman (2017), Rud et al. (2018). There is also a direct causal link from education to criminal behaviour; see Hjalmarsson et al. (2015).

[91] Oreopoulos and Salvanes (2011), Heckman et al. (2018).

[92] Conti et al. (2010).

3.4   Working Life and Parenthood

Individuals in their upper teens leaving the mandatory school system face a number of choices—start working or continue an educational career, engage in civil society at different levels, develop social relations as a preparation for family formation, and many others. These choices are conditioned on the prehistory, such as family background, cognitive and non-cognitive abilities acquired, and school performance.

Higher Education

The level of education reached is decisive for success in the labour market—in terms of income, work conditions and risk of unemployment.[93] Both the family background and the structure of inequality are important to educational decisions and thereby to educational mobility, as illustrated in Fig. 3.2.[94]

In all OECD countries, parental level of education is an important determinant of the educational choice of their children. Parents with higher education are more

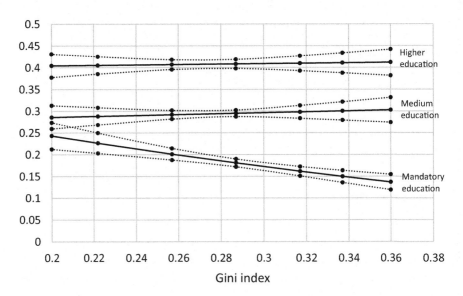

Fig. 3.2 Likelihood of going to higher education in OECD countries, depending on the Gini coefficient of the general disposable income distribution and the educational background of parents. The diagram illustrates the likelihood in question for three categories: children with at least one parent with higher education, children with at least one parent with medium-level education, and children whose parents have only the mandatory educational level. Uncertainty bands: average ± 1 std. Source: Cingano (2014)

---

[93] See, for example, OECD (2019a).

[94] Cingano (2014), Fig. 5.

likely to see their children go on to higher education. A high general level of income inequality in a society is associated with low mobility, but children of parents with a high or medium level of education are almost insensitive to the general level of inequality. Children whose parents have only the mandatory minimum level of education, by contrast, are highly sensitive; it is this category that explains most of the negative correlation between income inequality and educational mobility.

Not only the level of education attained but also skills are affected, including in this case among children with a medium educational background.

The most likely interpretation of this relationship is not that an ambitious system of income equalisation via taxes and transfers would automatically reduce the educational gap. A more probable causal link goes in the other direction—an educational system that also provides opportunities to children with a weak socio-economic and educational background will tend to increase educational mobility and equalise incomes.

The fact that the propensity to go on to higher education depends on family background even when children with the same cognitive capacities are compared is an old observation.[95] Both institutional and personal factors are involved. Free higher education and public loans on favourable terms lower the threshold into the higher education system.[96]

The choice between work and education is an example of the dilemma of deferred gratification,[97] which children with different personalities and prehistories are more or less apt to handle. Time preferences in adolescence are important for later outcomes not only concerning education but also income at middle age, unemployment, obesity and teen-age motherhood. It appears that early human capital investments are important to the degree of patience developed, whereas parental income and education as well as the cognitive ability of the child are less important in this respect.[98]

Across the world, females are increasing their educational level in relation to males.[99] In the long run, this can be expected to strengthen women's position in society, although there are many hurdles, political as well as economic.[100]

Beyond the choice to go to higher education lies the choice of branch. Even in the Nordic countries, which have achieved a relatively high degree of equality between the sexes,[101] professional choices tend to follow traditional lines, with implications also for equality in career possibilities and earnings.

---

[95] Erikson and Goldthorpe (1992), Bukodi et al. (2014).

[96] On the importance of education for inequality, see further Sect. 5.1.

[97] Mischel (2014) summarises several decades of research on this topic.

[98] Golsteyn et al. (2014).

[99] For a global overview, see World Bank (2018); for the OECD countries, see OECD (2019a).

[100] O'Malley (2010), World Bank (2018).

[101] OECD (2018b).

Working Life

In principle, a person's position in the labour market should be determined by abilities and effort, but narrowly defined skills such as *numeracy* and *literacy* are insufficient to explain differences in earnings, which can be considered the main indicator of labour market position.[102] Other abilities and resources contribute. *Non-cognitive abilities* are increasingly important.[103] In a study based on detailed Swedish data, a long-term increase in the returns to non-cognitive skills has been recorded.[104] It is particularly pronounced in the private sector and at the upper end of the wage distribution. Optimal skill mixes have apparently changed over time, and demand-side factors seem to be the main driving forces behind this development. Various explanations have been put forward, such as the economic maturation of the IT sector and the development of the global telecommunications infrastructure, which has facilitated offshoring of qualified services, both of which would increase competition with respect to purely cognitive abilities.[105]

A third component, beyond cognitive and non-cognitive skills, is *networks*. A well-developed network of social contacts improves matching in the labour market and is important both at the employee and employer level.[106]

Gaps in income, like cognitive abilities, tend to *widen with age* for one and the same cohort. This is a natural consequence of increasing differences in abilities and networks, and also in health (see below). As a consequence, long-term changes in income distribution can be affected by demographic changes. The pure time effect in a cohort may be difficult to disentangle from policy changes and the aggregate economic development. An increasing spread in earnings and consumption within cohorts has nonetheless been confirmed in several studies, the standard explanation being increasing human capital and cumulative advantage and disadvantage over the life cycle.[107]

Employment is not only a question of wages, however, but also of *security of income*. In recent decades, an increasing share of the workforce has worked under less stable relationships with an employer in the so-called "gig" or "crowd work" economy. Estimates vary depending on definitions but lie at around 10 per cent of the workforce in Europe and the United States.[108] For many, this is a supplemental source of income. It has some value as a way into the labour market—preferable to unemployment but less valuable than standard employment forms.[109] These new

---

[102] Castelló and Doménech (2002), Green et al. (2015), Devroye and Freeman (2001).

[103] Deming (2017).

[104] Edin et al. (2018).

[105] Beaudry et al. (2016) and Baldwin (2019), respectively.

[106] Granovetter (1973) and Montgomery (1991) are classical references. See Hensvik and Nordström Skans (2016) for a survey.

[107] Deaton and Paxson (1994), O'Rand (1996), Altonji et al. (2013), Crystal et al. (2017).

[108] Huws et al. (2016), Collins et al. (2019).

[109] Adermon and Hensvik (2020).

forms of employment raise a number of legal issues and problems concerning income security.[110]

Work environment is another factor of importance to health and well-being. Both physical disturbances in the form of pollution and the psychosocial environment have been studied in relation to socioeconomic status. In terms of pollution, the link is obvious, but also psychosocial environment tends to be more adverse for employees with a lower socioeconomic status, thereby reinforcing existing differences.[111] Inequality in health related to the psychosocial environment derives primarily from dimensions such as influence at work, possibilities for development, degrees of freedom, and meaning of work.

Discrimination

Two individuals with the same levels of cognitive and noncognitive abilities and similar networks are not always treated equally in the labour market, for reasons of discrimination with respect to gender, ethnicity or other personal characteristics.[112] Although there have been important advances with respect to equal treatment in some areas, the general picture is that patterns of discrimination are persistent.

Gender is one of the oldest and most stable bases for discrimination in both working life and civil society. At the global level, large differences between men and women still exist, and the female labour force participation is far below its potential, with important macroeconomic consequences.[113] Significant progress has been made in recent decades, but labour markets remain divided along gender lines, and significant wage differentials persist. Women also account for most of the unpaid work in society. Female representation in senior positions and in entrepreneurship remains low.

In the OECD area, the gap in annual average labour income between men and women has been reduced over a number of decades, but women's average annual labour income was still 39 per cent lower than that of men in 2015.[114] Income differentials persist also in the Nordic countries, in spite of conscious and systematic policies in the post-war period aimed at creating equal opportunities for both sexes.[115]

The gender gap in labour income can be related to three different factors—differences in employment rates, hours worked and hourly earnings. Although income differences exist already when young men and women enter the labour market, the gap is increased significantly when children are born.[116] This so-called *child penalty* has been the object of much research, both among economists and

---

[110] De Stefano (2016). On pension problems for this category, see OECD (2019b).

[111] For examples, see Kristensen et al. (2002) and Moncada et al. (2010) (Denmark and Spain). Kim and Cho (2020) provide an example that mental health problems due to adverse psychosocial environment may in fact be more pronounced in groups with a higher socioeconomic status.

[112] For an overview of research methods and results based on U.S. data, see Altonji and Blank (1999).

[113] Elborgh-Woytek et al. (2013).

[114] OECD (2018c).

[115] OECD (2018b).

[116] Angelov et al. (2016).

other social scientists.[117] There are clear similarities in the effects of child-bearing across countries but also sharp differences in the magnitude of the effects, due to differences in family policies (parental leave, childcare provision) and gender norms.

Other factors behind discriminatory behaviour include *ethnic origins*, various *handicaps*, and *age*.[118] Ethnic discrimination materialises for instance as a lower probability of being called for an interview when applying for a job.[119] Employer stereotypes about ability to learn new tasks, flexibility and ambition are important to age discriminatory behaviour. Signs of discrimination appear already at the age of 40.[120]

Family Life

A common pattern is that establishment in the labour market precedes family formation. A precarious position vis-à-vis the labour market may therefore lead to delayed or failed family formation,[121] an outcome that is particularly pronounced for men, given their traditional role in support of the family. School failures often lie behind such problems, whence young men with this background risk facing a triple dilemma—in school, in the labour market and in the marriage arena. As noted above, there is a long-term trend that women improve their educational level relative to men, a tendency that will aggravate this problem.

Once the family has been formed, more traditional forms of inequality develop. There is an existential inequality between parents and their children due to the total dependence of the latter on the former for their subsistence. In recent decades, steps have nonetheless been taken to redress some of this imbalance, for instance by the almost universal ratification of the Convention on the Rights of the Child and the legal prohibition of corporal punishment in a number of countries.

Patterns of inequality between parents have changed over a long period.[122] These behavioural changes are visible in the respective time budgets of parents. Women have increased their time in paid work and reduced time in unpaid activities, thereby strengthening their position both in the household and more generally. Men have increased their time in unpaid work, but not enough to compensate for women's reduction. Across Europe and the United States, there is evidence of convergence between the genders in unpaid work, but trends differ between housework and childcare. Gender convergence in housework has primarily resulted from women reducing their time, to some extent facilitated by increased productivity due to household machines. Childcare time has increased for both sexes, resulting in convergence only where men increased more than did women. These tendencies

---

[117] See Kleven et al. (2019) for a survey.

[118] There is a large literature on various bases for discrimination; for entries, see Eriksson et al. (2017).

[119] Bertrand and Mullainathan (2004) (U.S.), Carlsson and Rooth (2007) (Sweden).

[120] Carlsson and Eriksson (2019). For an overview of problems for the disabled in the labour market, see OECD (2009).

[121] Wolbers (2007).

[122] Pailhé et al. (2021).

are visible across countries with different family policies and gender norms, but the strength of convergence depends on the policy context.[123]

The changes in time use generated by the dual-earner norm to some extent depends on the level of education. Relatively more time is spent with children by parents with higher education.[124]

The persistent, uneven division of labour within the household, including in relatively egalitarian Nordic countries, explains why part-time work is most often the solution for the female parent. As noted above, the gender gap in wage earnings develops after the first child is born. Expectations about future engagement in the workplace play a role when employers decide on career possibilities, with consequences for hourly earnings.

Health

The link between socioeconomic status and health is well established.[125] On average, more advantaged individuals have better health. This link is important not only for those in poverty but at all levels of socioeconomic status. There are exceptions to the general rule that health status improves with education. Breast cancer, a common cancer form among women, is more prevalent among women with higher education.[126] The link is causal, not in the sense that studying engenders breast cancer, of course, but indirectly, because higher education changes the life course in a way that increases both the risk and the survival rate.

Some care should be exercised in the use of the term *socioeconomic status* in this context. It often covers diverse dimensions of well-being—education, income, social class or status—but correlations between these various interpretations and health outcomes differ.[127] For instance, there are clear effects on total death risk for all the factors mentioned except for income from work among women. For both men and women, there are clear links between education and mortality, and more precise statements can be made on specific death causes and ages.[128] Class and income show independent effects on mortality only among men, while status has effects only among women.

Causal links explaining these correlations could go in either direction. Some diseases acquired in childhood or youth could be sufficiently debilitating to lead to lower socioeconomic status, but they are generally too rare to account for the average association in adulthood in populations at large. The causal link in the other direction therefore dominates.

---

[123] Neilson and Stanfors (2014).

[124] Neilson and Stanfors (2018).

[125] Adler and Ostrove (1999), Marmot et al. (2012), Mackenbach et al. (2015).

[126] Palme and Simeonova (2015).

[127] Torssander and Erikson (2010).

[128] Lager and Torssander (2012) (Swedish data). For a wider perspective on rank and health, see Abbott et al. (2003) and Sapolsky (2004).

There are numerous mechanisms through which these causal links operate—ability to absorb information on health hazards, differences in work and living environments, and behavioural differences.

As for *health differences between the sexes*, independently of socioeconomic status, sick-leave absence displays the same pattern as earnings, namely small differences before the birth of the first child, then followed by a widening gap.[129] This indicates the presence of a medical factor in the development of careers and earnings over the life cycle. Traditional economic explanations for the uneven distribution of domestic work and the wider gender income gap in lower socioeconomic strata have focused on the lower wages of women, implying smaller household income losses when one of the parents is absent from work. But child-bearing implies physical and psychic stress in the pregnant woman, which can release a broad range of somatic and psychic disease conditions—pre-eclampsia, post-partum depression and others. Such effects, like other disease conditions, are particularly pronounced among lower socioeconomic strata and therefore contribute to income differences both between women and men and among women.[130]

Various forms of *addiction*—alcohol, tobacco, narcotic drugs, gaming—together represent an important health hazard. The consumption profiles across socioeconomic strata differ. The use of tobacco decreases with increasing socioeconomic status, whereas gambling shows the reverse pattern (in absolute terms), and consumption of alcohol and narcotic drugs (at least in moderate quantities) is more independent of socioeconomic status. The capacity to deal with addiction problems varies, however.

Consumer Perspectives
The relationship between consumers and producers is unavoidably asymmetric. The producer always knows more about the goods and services produced than the consumer. Different consumers are more or less well equipped to handle this asymmetry, which is yet another source of inequality.

These differences may affect both strategic and small-scale economic decisions. Loans and savings are important to the household economy and also have long-term implications, for instance for decisions on retirement. *Financial literacy* is a key asset in this context. As expected, there is a strong link between socioeconomic background and financial literacy, both among young people and adults.[131]

All human beings are worse decision-makers when under *stress*.[132] But low-income households run a greater risk of ending up in stressful situations, and the relative cost of making bad decisions is greater for this category. Applying even for a small short-term loan at high interest may lead to a poverty trap—a permanent situation of household deficit and lifelong economic distress.

---

[129] Angelov et al. (2020).

[130] Fransson et al. (2021).

[131] Lusardi et al. (2010), Lusardi and Mitchell (2014).

[132] Shah et al. (2012), Mani et al. (2013), Mullainathan and Shafir (2013).

3.5   Old Age

Living standards and longevity among retired are to a considerable extent determined by the earlier stages of the life cycle. The effects of factors that contribute to success and health in these earlier stages—education, income, favourable living and work environments—become visible. By way of example, the (raw) gap in life expectancy across OECD countries between highly educated persons and persons with only mandatory education is 3.5 years for men and 2.5 years for women at the age of 65, corresponding to around 15 per cent of the remaining life expectancy.[133]

This is an area where the distinction between absolute and relative measures becomes particularly important. The period after retirement is a limited part of life, and also small absolute differences in longevity can make a large difference in relative terms.

Life expectancy has increased steadily in OECD countries for a long time, and increasing longevity has almost acquired the status of a natural law. As noted earlier, however, within the European Union, member countries differ in this respect, and in the United States inequality in longevity has risen.[134] In Sweden, life expectancy for women with a short education at the age of 30 decreased by 0.2 years between 2012 and 2017.[135]

The life expectancy measure can be refined using quality-adjusted life years (QALYs), which measure the degree to which quality of life is impaired by ill-health. For example, the distribution of QALYs in Sweden became more uneven during the last two decades of the twentieth century.[136]

Income sources for persons over 65 include work, public pensions, private pensions and private wealth. Within the OECD countries, public pensions dominate on average. The expansion of public pension systems during the twentieth century has reduced the risk of poverty among retirees, but the poverty rate among persons older than 65 is still higher at 13.5 per cent compared to 11.8.[137] Inequality among retirees is on average lower than in the population; the Gini coefficient across OECD is 0.302, compared to 0.317 for the total population.[138] The public pension systems are the main equalising force; inequality is higher in countries where private pension schemes are more important.[139] But pension disparities between men and women are large, 25 per cent on average, with particularly large differences in Germany and the Netherlands.[140] Average inequality across the OECD within the age category of

---

[133] Murtin et al. (2017). Refers to 23 OECD countries around 2011.

[134] Debón Aucejo et al. (2017) (EU), Bosworth et al. (2016) (U.S.).

[135] Statistics Sweden (2018). This figure comprises only women born in Sweden and is not a composition effect.

[136] Burström et al. (2005).

[137] OECD (2019b). Figures refer to percentage with income lower than 50% of median equivalised household disposable income.

[138] OECD (2019b).

[139] Been et al. (2016).

[140] OECD (2019b). For a deeper analysis, see Lis and Bonthuis (2019).

65 and above as measured by the Gini coefficient has been stable in recent decades, but variations are large; Sweden and New Zealand have seen particularly large increases.

3.6 Summary

The research cited on mechanisms generating inequality at the individual level can be summarised with the following list of stylised facts:

- Human well-being depends on a number of different dimensions—education, health, income and others. There is significant co-variation between these different dimensions, but causal links can go in either direction and have only been partly mapped and explained.
- The distinction between opportunities and outcomes, commonly used in debates, is difficult to maintain in practice. Outcomes in one stage of human life set the initial conditions for the next stage.
- The resources at the disposal of the individual at a given point in time affect the ability to profit from opportunities and to cope with difficulties. Such self-reinforcing mechanisms exist in a number of areas; success begets success.
- The age-old opposition of genes versus environment has been made obsolete by modern research on genetics and epigenetics. The development of the individual is the result of a complex interaction between genes and the environment that starts with conception, or even before, and continues into adulthood. Some of the epigenetic effects are irreversible, and some are even heritable. In general terms, environmental influence is more important than acknowledged in the bulk of twentieth-century literature.
- The individual life course is frontloaded. Family background and early phases—prenatal, preschool and school—have a decisive impact on later life outcomes in a number of dimensions. The probability of children from different socioeconomic backgrounds going on to higher education differs even among children with the same levels of ability.
- Gaps in cognitive abilities, health and income tend to widen with age for one and the same cohort.
- Individuals are not rewarded according to their abilities in the labour market. Discrimination with respect to gender, ethnic background and other arbitrary characteristics persists in spite of advances in equal treatment in recent decades.

These facts are the building blocks of formal models to be treated in the following chapter.

Bibliography

Abbott, D. H., et al. (2003). Are subordinates always stressed? A comparative analysis of rank differences in cortisol levels among primates. *Hormones and Behavior, 43*, 67–82.

Adermon, A., & Hensvik, L. (2020). Gig-jobs: Stepping stones or dead ends? *IFAU Working Paper* 2020:23. Uppsala: IFAU.

Adermon, A., et al. (2018). Intergenerational wealth mobility and the role of inheritance: Evidence from multiple generations. *The Economic Journal, 128*(612), F482–F513.

Adler, N., & Ostrove, J. (1999). SES and health: What we know and what we don't. *Annals of the New York Academy of Sciences, 896*, 3–15.

Aizer, A., et al. (2016). Maternal stress and child outcomes: Evidence from siblings. *Journal of Human Resources, 51*(3), 523–555.

Alati, R., et al. (2006). In utero alcohol exposure and prediction of alcohol disorders in early adulthood. *Archives of General Psychiatry, 63*, 1009–1016.

Alesina, A., et al. (2021). Intergenerational mobility in Africa. *Econometrica, 89*(1), 1–35.

Alessie, R., et al. (2019). Economic conditions at birth and cardiovascular disease risk in adulthood: Evidence from post-1950 cohorts. *Social Science and Medicine, 224*, 77–84.

Almond, D., & Currie, J. (2011). Human capital development before age five. In D. Card & O. Ashenfelter (Eds.), *Handbook of labor economics* (Vol. 4B). Elsevier.

Almond, D., & Mazumder, B. (2005). The 1918 influenza pandemic and subsequent health outcomes: An analysis of SIPP data. *American Economic Review, 95*(2), 258–262.

Almond, D., & Mazumder, B. (2013). Fetal origins and parental responses. *Annual Review of Economics, 5*, 37–56.

Almond, D., et al. (2009). Chernobyl's sub-clinical legacy: Prenatal exposure to radioactive fallout and school outcomes in Sweden. *Quarterly Journal of Economics, 124*(4), 1729–1772.

Almond, D., et al. (2014). In utero Ramadan exposure and children's academic performance. *The Economic Journal, 125*, 1501–1533.

Almond, D., et al. (2018). Childhood circumstances and adult outcomes: Act II. *Journal of Economic Literature, 56*(4), 1360–1446.

Alsan, M., & Goldin, C. (2015). Watersheds in infant mortality: The role of effective water and sewerage infrastructure, 1880 to 1915. *NBER working paper* 21263. Cambridge, MA: National Bureau of Economic Research.

Altonji, J. G., & Blank, R. M. (1999). Race and gender in the labor market. Ch. 48 in O. Ashenfelter, & D. Card (Eds.), *Handbook of labor economics*, Vol. 3. Amsterdam: Elsevier.

Altonji, J., et al. (2013). Modeling earnings dynamics. *Econometrica, 81*(4), 1395–1454.

Andersen, I. G., & Andersen, S. C. (2015). Student-centered instruction and academic achievement: Linking mechanisms of educational inequality to schools' instructional strategy. *British Journal of Sociology of Education, 38*(4), 533–550.

Andreasson, S., et al. (2020). Alkohol, graviditet och spädbarns hälsa – ett gemensamt ansvar. *Alkoholen och samhället 2020* (Alcohol, pregnancy and baby health – a common concern. Alcohol and Society 2020). Stockholm: Swedish Association of Nurses et al.

Angelov, N., et al. (2016). Is the persistent gender gap in income and wages due to unequal family responsibilities? *Journal of Labor Economics, 34*(3), 545–579.

Angelov, N., et al. (2020). Sick of family responsibilities? *Empirical Economics, 58*, 777–814.

Arima, Y., & Fukuoka, H. (2019). Developmental origins of health and disease theory in cardiology. *Journal of Cardiology, 76*(1), 14–17.

Bäckman, O. (2017). High school dropout, resource attainment, and criminal convictions. *Journal of Research in Crime and Delinquency, 54*(5), 715–749.

Baldwin, R. (2019). *The globotics upheaval. Globalization, robotics and the future of work.* Weidenfeld & Nicholson.

Beaudry, P., et al. (2016). The great reversal in the demand for skill and cognitive tasks. *Journal of Labor Economics, 34*(1), S199–S247.

Been, J., et al. (2016). Public/private pension mix, income inequality and poverty among the elderly in Europe: An empirical analysis using new and revised OECD data. *Social Policy and Administration, 51*(7), 1079–1100.

Behnke, M., & Smith, V. C. (2013). Prenatal substance abuse: Short- and long-term effects on the exposed fetus. *Pediatrics, 2013*, e1009–e1025.

Bengtsson, T., & van Poppel, F. (2011). Socioeconomic inequalities in death from past to present: An introduction. *Explorations in Economic History, 48*(3), 343–356.

Bertrand, M., & Mullainathan, S. (2004). Are Emily and Greg more employable than Lakisha and Jamal? A field experiment on labor market discrimination. *American Economic Review, 94*(4), 991–1013.

Bjerk, D. (2012). Re-examining the impact of dropping out on criminal and labor outcomes in early adulthood. *Economics of Education Review, 31*(1), 110–122.

Björklund, A., & Jäntti, M. (1997). Intergenerational income mobility in Sweden compared to the United States. *American Economic Review, 87*(5), 1009–1018.

Björklund, A., & Jäntti, M. (2009). Intergenerational income mobility and the role of family background. In W. Salverda et al. (Eds.), *Oxford handbook of economic inequality* (pp. 491–521). Oxford University Press.

Björklund, A., et al. (2012). Intergenerational top income mobility in Sweden: Capitalist dynasties in the land of equal opportunity? *Journal of Public Economics, 96*, 474–484.

Black, S., et al. (2019). This is only a test? Long-run and intergenerational impacts of prenatal exposure to radioactive fallout. *Review of Economics and Statistics, 101*(3), 531–546.

Black, S., et al. (2020). Poor little rich kids? The role of nature versus nurture in wealth and other economic outcomes and behaviours. *The Review of Economic Studies, 87*(4), 1683–1725.

Blanden, J. (2009). How much can we learn from international comparisons of intergenerational mobility? *Discussion Paper* No. 111, *Centre for the Economics of Education, London School of Economics and Political Science*.

Borgerhoff Mulder, M. et al. (2009). Intergenerational wealth transmission and the dynamics of inequality in small-scale societies. *Science* (October 30) 326(5953), 682–688.

Bornehag, C.-G., et al. (2015). Prenatal phthalate exposures and anogenital distance in Swedish boys. *Environmental Health Perspectives*. https://doi.org/10.1289/ehp.1408163.

Bosworth, B. et al. (2016). Later retirement, inequality in old age, and the growing gap in longevity between rich and poor. *Brookings Economic Studies*. Washington, DC: The Brookings Institution.

Bowles, S., et al. (2010). Intergenerational wealth transmission and inequality in premodern societies. Special issue of *Current Anthropology, 51*(1).

Bradley, R. H., & Corwyn, R. F. (2002). Socioeconomic status and child development. *Annual Review of Psychology, 53*, 371–399.

Brandén, G. (2019). Does inequality reduce mobility? The Great Gatsby curve and its mechanisms. *IFAU Working Paper* 2019:20, Uppsala.

Brunori, P., et al. (2013). Inequality of opportunity, income inequality, and economic mobility: Some international comparisons. In E. Paus (Ed.), *Getting development right*. Palgrave Macmillan.

Bukodi, E., et al. (2014). The effects of social origins and cognitive ability on educational attainment: Evidence from Britain and Sweden. *Acta Sociologica, 57*(4), 293–310.

Burger, K. (2010). How does early childhood care and education affect cognitive development? An international review of the effects of early interventions for children from different social backgrounds. *Early Childhood Research Quarterly, 25*, 140–165.

Burström, K., et al. (2005). Increasing socio-economic inequalities in life expectancy and QALYs in Sweden 1980–1997. *Health Economics, 14*(8), 831–850.

Carlsson, M., & Eriksson, S. (2019). Age discrimination in hiring decisions: Evidence from a field experiment in the labor market. *Labour Economics, 59*, 173–183.

Carlsson, M., & Rooth, D.-O. (2007). Evidence of ethnic discrimination in the Swedish labor market using experimental data. *Labour Economics, 14*(4), 716–729.

Case, A., et al. (2002). Economic status and health in childhood: The origins of the gradient. *American Economic Review, 92*(5), 1308–1334.

Castelló, A., & Doménech, R. (2002). Human capital inequality and economic growth: Some new evidence. *The Economic Journal, 112*, C187–C200.

Chall, J. (2000). *The academic achievement challenge: What really works in the classroom.* Guilford.

Chaudhury, N., et al. (2006). Missing in action: Teacher and health worker absence in developing countries. *Journal of Economic Perspectives, 20*(1), 91–116.

Chen, Y., & Li, H. (2009). Mother's education and child health: Is there a nurturing effect? *Journal of Health Economics, 28*(2), 413–426.

Chetty, R., et al. (2014). Where is the land of opportunity? The geography of intergenerational mobility in the United States. *The Quarterly Journal of Economics, 129*(4), 1553–1623.

Cingano, F. (2014). Trends in income inequality and its impact on economic growth. *OECD social, employment and migration working papers*, No. 163, OECD Publishing.

Claussen, B., et al. (2003). Impact of childhood and adulthood socioeconomic position on cause specific mortality: The Oslo Mortality Study. *Journal of Epidemiology and Community Health, 57*, 40–45.

Cnattingius, S. (2004). The epidemiology of smoking during pregnancy: Smoking prevalence, maternal characteristics, and pregnancy outcomes. *Nicotine & Tobacco Research, 6*, Suppl. 2 (April 2004), S125–S140.

Cole, S. W., et al. (2007). Social regulation of gene expression in human leukocytes. *Genome Biology, 8*, R189. https://doi.org/10.1186/gb-2007-8-9-r189

Cole, S. W. et al. (2012). Transcriptional modulation of the developing immune system by early life social adversity. *Proceedings of the National Academy of Sciences, 109*(50), 20578–20583 (December 11, 2012).

Collado, M. D., et al. (2020). *Estimating intergenerational and assortative processes in extended family data.* Working paper, submitted for publication. Accessible at https://janstuhler.files.wordpress.com/2020/10/collado-ortuno-and-stuhler-manucript_v37.pdf.

Collins, B. et al. (2019). Is gig work replacing traditional employment? Evidence from two decades of tax returns. *Working paper, internal revenue service and Stanford University.* SOI Working Paper March 2019.

Conti, G., et al. (2010). The education-health gradient. *American Economic Review, 100*(2), 234–238.

Conti, G., et al. (2012). Primate evidence on the late health effects of early-life adversity. *Proceedings of the National Academy of Sciences, 109*(23), 8866–8871.

Corak, M. (2013). Income inequality, equality of opportunity, and intergenerational mobility. *Journal of Economic Perspectives, 27*(3), 79–102.

Corak, M. (2017). Divided landscapes of economic opportunity: The Canadian geography of intergenerational income mobility. *HCEO working paper* No. 2017–043, University of Chicago.

Crystal, S., et al. (2017). Cumulative advantage, cumulative disadvantage, and evolving patters of late-life inequality. *The Gerontologist, 57*(5), 910–920.

Cunha, F., & Heckman, J. J. (2009). The economics and psychology of inequality and human development. *Journal of the European Economic Association, 7*(2–3), 320–364.

Cunha, F., et al. (2006). Interpreting the evidence on life cycle skill formation. Chapter 12 in *Handbook of the economics of education*, vol. 1, 697–812. : North-Holland.

Currie, J. (2009). Healthy, wealthy, and wise: Socioeconomic status, poor health in childhood, and human capital development. *Journal of Economic Literature, 47*(1), 87–122.

Currie, J. (2013). Pollution and infant health. *Child Development Perspectives, 7*(4), 237–242.

Currie, J., & Moretti, E. (2007). Biology as destiny? Short- and long-run determinants of intergenerational transmission of birth weight. *Journal of Labour Economics, 25*(2), 231–264.

Currie, J., & Rossin-Slater, M. (2015). Early-life origins of lifecycle well-being: Research and policy implications. *Journal of Policy Analysis and Management, 34*(1), 208–242.

Cutler, D. M., & Lleras-Muney, A. (2010). Understanding differences in health behaviors by education. *Journal of Health Economics, 29*(1), 1–28.

Cutler, D. M., & Miller, G. (2005). The role of public health improvements in health advances: The 20th century United States. *NBER working paper* 10511. Cambridge, MA: National Bureau of Economic Research.

Cutler, D. M., et al. (2006). The determinants of mortality. *Journal of Economic Perspectives, 20*(3), 97–120.

De Nardi, M. (2004). Wealth inequality and intergenerational links. *The Review of Economic Studies, 71*(3), 743–768.

De Stefano, V. (2016). *The rise of the «just-in-time workforce»: On-demand work, crowdwork and labour protection in the «gig-economy».* International Labour Office.

Deaton, A., & Paxson, C. (1994). Intergenerational choice and inequality. *Journal of Political Economy, 102*(3), 437–467.

Debón Aucejo, A. M., et al. (2017). Characterization of between-group inequality of longevity in European Union countries. *Insurance Mathematics and Economics, 75*, 151–165.

Dedman, D. J., et al. (2001). Childhood housing conditions and later mortality in the Boyd Orr cohort. *Journal of Epidemiology and Community Health, 55*, 10–15.

Deming, D. J. (2017). The growing importance of social skills in the labor market. *The Quarterly Journal of Economics, 132*(4), 1593–1640.

Desai, S. B., et al. (2010). *Human development in India: Challenges for a society in transition.* Oxford University Press.

Devroye, D., & Freeman, R. B. (2001). Does inequality in skills explain inequality in earnings across advanced countries? *NBER working paper* No. 8140. Cambridge, MA: National Bureau of Economic Research.

Dietrichson, J., et al. (2020). Universal preschool programs and long-term child outcomes: A systematic review. *Journal of Economic Surveys, 34*(5), 1007–1043.

Drèze, J., & Sen, A. (2013). *An uncertain glory. India and its contradictions.* Allen Lane.

Durlauf, S. N., & Seshadri, A. (2017). Understanding the great gatsby curve. Ch. 4 in *NBER Macroeconomics annual 2017*, 333–393. Cambridge, MA: National Bureau of Economic Research.

Edin, P.-A., et al. (2018). The rising return to non-cognitive skill. *IFAU working paper* 2018:18. Uppsala: IFAU.

Einarson, A., & Riordan, S. (2009). Smoking during pregnancy and lactation: A review of risks and cessation strategies. *European Journal of Clinical Pharmacology, 65*, 325–330.

Elborgh-Woytek, K., et al. (2013). Women, work, and the economy: Macroeconomic gains from gender equity. *IMF Staff discussion note* 13/10. Washington, DC: The International Monetary Fund.

Erikson, R., & Goldthorpe, J. H. (1992). *The constant flux: A study of class mobility in industrial societies.* Clarendon Press.

Eriksson, S., et al. (2017). What is the right profile for getting a job? *Empirical Economics, 53*(2), 803–826.

Ermisch, J., et al. (Eds.). (2012). *From parents to children. The intergenerational transmission of advantage.* Russell Sage Foundation.

Evans, G. W., & Schamberg, M. A. (2009). Childhood poverty, chronic stress, and adult working memory. *Proceedings of the National Academy of Sciences* (April 21, 2009) 106(16), 6545–6549.

Fan, Y. et al. (2018). The Great Gatsby curve in China: Economic transition, inequality and intergenerational mobility. *Working paper, Chinese University of Hong Kong* (Hong Kong,).

Fogel, R. W. (1986). Nutrition and the decline in mortality since 1700: Some preliminary findings. Ch. 9 in S. L. Engerman, & R. E. Gallman (Eds.), *Long-term factors in American economic growth.* Chicago: University of Chicago Press.

Fogel, R. W., et al. (1983). Secular changes in American and British stature and nutrition. *The Journal of Interdisciplinary History, 14*(2), 445–481.

Fransson, E., et al. (2021). *Kvinnors hälsa, sjukfrånvaro och inkomster efter barnafödande* (Female health, sick-leave absence and income post family formation). *IFAU Report 2021:17*, Uppsala.

Fratiglioni, L., et al. (2020). Ageing without dementia: Can stimulating psychosocial and lifestyle experiences make a difference? *Lancet Neurology, 19*, 533–543.

Gallardo-Albarrán, D. (2020). Sanitary infrastructures and the decline of mortality in Germany, 1877–1913. *Economic History Review, 73*(3), 730–757.

Galster, G., et al. (2007). The influence of neighborhood poverty during childhood on fertility, education, and earnings outcomes. *Housing Studies, 22*(5), 723–751.

Golsteyn, B. H. H., et al. (2014). Adolescent time preferences predict lifetime outcomes. *The Economic Journal, 124*(580), F739–F761.

Graff Zivin, J., & Neidell, M. (2013). Environment, health, and human capital. *Journal of Economic Literature, 51*(3), 689–730.

Graignic-Philippe, R., et al. (2014). Effects of prenatal stress on fetal and child development: A critical literature review. *Neuroscience & Biobehavioral Reviews, 43*, 137–162.

Granovetter, M. S. (1973). The strength of weak ties. *American Journal of Sociology, 78*(6), 1360–1380.

Green, A., et al. (2015). Cross-country variation in adult skills inequality: Why are skill levels and opportunities so unequal in anglophone countries? *Comparative Education Review, 59*(4), 595–618.

Grönqvist, H., et al. (2020). Understanding how low levels of early lead exposure affect children's life trajectories. *Journal of Political Economy, 128*(9), 3376–3433.

Grossman, M. (2008). The relationship between health and schooling. *Eastern Economic Journal, 34*(3), 281–292.

Güell, M., et al. (2018). Correlating social mobility and economic outcomes. *Economic Journal, 128*(612), F353–F403.

Gustafsson, J.-E., et al. (2010). *School, learning and mental health. A systematic review*. The Royal Swedish Academy of Sciences.

Haeck, C., et al. (2012). The distributional impacts of a universal school reform on mathematical achievements: A natural experiment from Canada. *Working Paper, Centre Inter-universitaire sur le Risque, les Politiques Économiques et d'Emploi*. Montréal: Université du Québec.

Halliday, T., et al. (2018). Intergenerational health mobility in the US. *IZA Working paper* 11304, .

Hanushek, E., et al. (2015). Returns to skills around the world: Evidence from PIAAC. *European Economic Review, 73*, 103–130.

Harms, M. B., et al. (2017). Early life stress, FKBP5 methylation, and inhibition-related prefrontal function: A prospective longitudinal study. *Development and Psychopathology, 29*(5), 1895–1903.

Hart, B., & Risley, T. R. (2003). The early catastrophe. The 30 million word gap. *American Educator, 27*(1), 4–9.

Heckman, J. J. (2008). Schools, skills and synapses. *Economic Inquiry, 46*(3), 289–324.

Heckman, J. J., et al. (2018). Returns to education: The causal effects of education on earnings, health, and smoking. *Journal of Political Economy, 126*(S1), S197–S246.

Hedman, L., et al. (2015). Cumulative exposure to disadvantage and the intergenerational transmission of neighbourhood effects. *Journal of Economic Geography, 15*, 195–215.

Heidrich, S. (2017). Intergenerational mobility in Sweden: A regional perspective. *Journal of Population Economics, 30*, 1241–1280.

Helgertz, J., & Önnerfors, M. (2019). Public water and sewerage investments and the urban mortality decline in Sweden 1875–1930. *The History of the Family, 24*(2), 307–338.

Hensch, T. K., & Bilimoria, P. M. (2012). Re-opening windows: Manipulating critical periods for brain development. *Cerebrum, 2012*, 11.

Hensvik, L., & Nordström Skans, O. (2016). Social networks, employee selection, and labor market outcomes. *Journal of Labor Economics, 34*(4), 825–867.

Hirsch, E. D., Jr. (2016). *Why knowledge matters. Rescuing our children from failed educational theories*. Harvard Education Press.

Hjalmarsson, R., et al. (2015). The effect of education on criminal convictions and incarceration: Causal evidence from micro-data. *The Economic Journal, 125*, 1290–1326.

Hoff, E. (2006). How social contexts support and shape language development. *Developmental Review, 26*, 55–88.

Hoff, E., et al. (2002). Socioeconomic status and parenting. In Bornstein, M. (Ed.), *Handbook of parenting*, 2nd ed., Vol. 2. Biology and ecology of parenting. Mahwah, NJ: Lawrence Erlbaum Associates, Publishers.

Hoffman, D. J., et al. (2017). Developmental origins of health and disease: Current knowledge and potential mechanisms. *Nutrition Reviews, 75*(12), 951–970.

Huizink, A. C. (2014). Prenatal cannabis exposure and infant outcomes: Overview of studies. *Progress in Neuro-Psycho-pharmacology & Biological Psychiatry, 52*, 45–52.

Huws, U., et al. (2016). *Crowd work in Europe: Preliminary results from a survey in the UK, Sweden, Germany, Austria and the Netherlands*. First Draft Report, FEPS/UNI-Europa. University of Hertfordshire.

Jablonka, E., & Lamb, M. J. (2020). *Inheritance systems and the extended synthesis*. Cambridge University Press.

Jablonka, E., & Raz. (2009). Transgenerational epigenetic inheritance: Prevalence, mechanisms, and implications for the study of heredity and evolution. *The Quarterly Review of Biology, 84*(2), 131–176.

Jans, J., et al. (2018). Economic status, air quality, and child health: Evidence from inversion episodes. *Journal of Health Economics, 61*, 220–232.

Jäntti, M., & Jenkins, S. P. (2015). Income mobility. Chapter 10 in A. B. Atkinson, & F. Bourguignon (Eds.), *Handbook of income distribution*, Vol. 2, 807–935.

Jürges, H. (2015). Ramadan fasting, sex-ratio at birth, and birth weight: No effects on Muslim infants born in Germany. *Economics Letters, 137*, 13–16.

Kesztenbaum, L., & Rosenthal, J.-L. (2017). Sewers' diffusion and the decline of mortality: The case of Paris, 1880–1914. *Journal of Urban Economics, 98*, 174–186.

Kim, Y.-M., & Cho, S.-I. (2020). Socioeconomic status, work-life conflict, and mental health. *American Journal of Industrial Medicine, 63*(8), 703–712.

Kleven, H., et al. (2019). Child penalties across countries: Evidence and explanations. *AEA Papers and Proceedings, 109*(May), 122–126.

Kristensen, T. S., et al. (2002). Socioeconomic status and psychosocial work environment: Results from a Danish national study. *Scandinavian Journal of Public Health, 30*, 41–48.

Lager, A. C. J., & Torssander, J. (2012). Causal effect of education on mortality in a quasi-experiment on 1.2 million Swedes. *Proceedings of the National Academy of Sciences*, May 29, 2012, 109(22), 8461–8466.

Lareau, A. (2002). Social class and childrearing in black families and white families. *American Sociological Review, 67*, 747–776.

Lee, C. (2014). In utero exposure to the Korean War and its long-term effects on socioeconomic and health outcomes. *Journal of Health Economics, 33*, 76–93.

Lee, J. J., et al. (2018). Gene discovery and polygenic prediction from a genome-wide association study of educational attainment in 1.1 million individuals. *Nature Genetics, 50*, 1112–1121.

Lis, M., & Bonthuis, B. (2019). Drivers of the gender gap in pensions: Evidence from EU-SILC and the OECD pension model. In Holzmann, R. et al. (Eds.), *Progress and challenges of nonfinancial defined contribution pension schemes: Volume 2. Addressing gender, administration, and communication*. Washington, DC: The World Bank.

López Turley, R. N. (2003). Are children of young mothers disadvantaged because of their mother's age or family background? *Child Development, 74*(2), 465–474.

Lumey, L. H., et al. (2011). Prenatal famine and adult health. *Annual Review of Public Health, 32*, 237–262.

Lusardi, A., & Mitchell, O. S. (2014). The economic importance of financial literacy: Theory and evidence. *Journal of Economic Literature, 52*(1), 5–44.

Lusardi, A., et al. (2010). Financial literacy among the young. *Journal of Consumer Affairs, 44*(2), 358–380.

Mackenbach, J. P., et al. (2015). Variations in the relation between education and cause-specific mortality in 19 European populations: A test of the "fundamental causes" theory of social inequalities in health. *Social Science & Medicine, 127*, 51–62.

Majid, M. F. (2015). The persistent effects of in utero nutrition shocks over the life cycle: Evidence from Ramadan fasting. *Journal of Development Economics, 117*, 48–57.

Mani, A., et al. (2013). Poverty impedes cognitive function. *Science, 341*, 976–980.

Marmot, M., et al. (2012). WHO European review of social determinants of health and the health divide. *The Lancet, 380*, 1011–1029.

May, P. A., et al. (2018). Prevalence of fetal alcohol spectrum disorders in 4 US communities. *Journal of the American Medical Association, 319*(5), 474–482.

Mayer, R. E. (2004). Should there be a three-strikes rule against pure discovery learning? The case for guided methods of instruction. *American Psychologist, 59*(1), 14–19.

McCall, S. J., et al. (2015). Evaluating the social determinants of teenage pregnancy: A temporal analysis using a UK obstetric database from 1950 to 2010. *Journal of Epidemiology and Community Health, 15*(69), 49–54.

McKeown, T. (1976). *The modern rise of population.* Academic.

Mischel, W. (2014). *The marshmallow test: Why self-control is the engine of success.* Little, Brown & Co.

Mohammed, A. H., et al. (2002). Environmental enrichment and the brain. Chapter 8 in M. A. Hofman et al. (Eds.), *Progress in brain research*, vol. 138. Amsterdam: Elsevier.

Moncada, S., et al. (2010). Psychosocial work environment and its association with socioeconomic status. A comparison of Spain and Denmark. *Scandinavian Journal of Public Health, 38*(Suppl 3), 137–148.

Montgomery, J. (1991). Social networks and labor-market outcomes: Toward an economic analysis. *The American Economic Review, 81*(5), 1408–1418.

Moore, D. S. (2015). *The developing genome. An introduction to behavioral genetics.* Oxford University Press.

Mullainathan, S., & Shafir, E. (2013). *Scarcity. The true cost of not having enough.* Picador, Henry Holt and Company.

Murtin, F., et al. (2017). Inequalities in longevity by education in OECD countries: Insights from new OECD estimates. *OECD statistics working papers* 2017/02. Paris: OECD.

Musset, P. (2012). School choice and equity: Current policies in OECD countries and a literature review. *Directorate for education working paper* N°66. Paris: OECD.

Narayan, A., et al. (2018). *Fair progress? Economic mobility across generations around the world.* World Bank.

Neilson, J., & Stanfors, M. (2014). It's about time! Gender, parenthood, and household divisions of labor under different welfare regimes. *Journal of Family Issues, 35*(8), 1066–1088.

Neilson, J., & Stanfors, M. (2018). Time alone or together? Trends and trade-offs among dual-earner couples, Sweden 1990–2010. *Journal of Marriage and Family, 80*, 80–98.

Nettle, D., et al. (2011). Early-life conditions and age at first pregnancy in British women. *Proceedings of the Royal Society B, 278*, 1721–1727.

Nilsson, J. P. (2017). Alcohol availability, prenatal conditions, and long-term economic outcomes. *Journal of Political Economy, 125*(4), 1149–1207.

Norman, R. E., et al. (2012). The long-term health consequences of child physical abuse, emotional abuse, and neglect: A systematic review and meta-analysis. *Public Library of Science Medicine, 9*(11), e1001349.

Nybom, M. (2018). *Intergenerational mobility: A dream deferred?* International Labour Office.

O'Malley, B. (2010). *Education under attack.* Unesco.

O'Rand, A. M. (1996). The precious and the precocious: Understanding cumulative disadvantage and cumulative advantage over the life course. *The Gerontologist, 36*(2), 230–238.

OECD. (2009). *Sickness, disability and work: Keeping on track in the economic downturn.* Background paper for Stockholm conference 14–15 May, 2009. Paris: OECD.

OECD. (2016). *PISA 2015 results (volume II): Policies and practices for successful schools.* PISA, OECD Publishing, Paris.

OECD. (2017). *Educational opportunity for all. Overcoming inequality throughout the life course.* Paris: OECD Publications.

OECD. (2018a). How do science teachers teach science – and does it matter? *PISA in Focus* 2018#90. Paris: OECD Publishing.

OECD. (2018b). *Is the last mile the longest? Economic gains from gender equality in the Nordic countries.* Paris: OECD Publishing.

OECD. (2018c). *OECD employment outlook 2018.* Paris: OECD Publishing.

OECD. (2019a). *Education at a glance 2019: OECD indicators.* OECD Publishing.

OECD. (2019b). *Pensions at a glance 2019: OECD and G20 indicators.* OECD Publishing.

Oosterbeek, H., & van der Klaauw, B. (2013). Ramadan, fasting and educational outcomes. *Economics of Education Review, 34,* 219–226.

Oreopoulos, P., & Salvanes, K. G. (2011). Priceless: The non-pecuniary benefits of schooling. *Journal of Economic Perspectives, 25*(1), 159–184.

Pailhé, A., et al. (2021). The great convergence: Gender and unpaid work in Europe and the United States. *Population and Development Review,* 08 February 2021. https://doi.org/10.1111/padr.12385.

Palme, M., & Simeonova, E. (2015). Does women's education affect breast cancer risk and survival? Evidence from a population based social experiment in education. *Journal of Health Economics, 42,* 115–124.

Persson, P., & Rossin-Slater, M. (2018). Family ruptures, stress, and the mental health of the next generation. *American Economic Review, 108*(4–5), 1214–1252.

Petronis, A. (2010). Epigenetics as a unifying principle in the aetiology of complex traits and diseases. *Nature, 465,* 721–727.

Piketty, T. (2000). Theories of persistent inequality and intergenerational transmission. In A. B. Atkinson & F. Bourguignon (Eds.), *Handbook of income distribution* (Vol. 1). North-Holland.

Piketty, T., & Zucman, G. (2015). Wealth and inheritance in the long run. Ch. 15 in A. B. Atkinson, & F. Bourguignon (Eds.), *Handbook of income distribution,* Vol. 2, pp. 1303–1368.

Preston, S. H. (1975). The changing relation between mortality and level of economic development. *Population Studies, 29*(2), 231–248.

Prus, S. G. (2007). Age, SES, and health: A population level analysis of health inequalities over the lifecourse. *Sociology of Health & Illness, 29*(2), 275–296.

Reeves, R. V., & Krause, E. (2019). The effects of maternal depression on early childhood development and implications for economic mobility. *Brookings Economic Studies.* Washington, DC: The Brookings Institution.

Rud, I., et al. (2018). What drives the relationship between early criminal involvement and school dropout? *Journal of Quantitative Criminology, 34,* 139–166.

Sanders, N. (2012). What doesn't kill you makes you weaker. Prenatal pollution exposure and educational outcomes. *Journal of Human Resources, 47*(3), 826–850.

Sandewall, Ö., et al. (2014). The co-twin methodology and returns to schooling—testing a critical assumption. *Labor Economics, 26,* 1–10.

Sapolsky, R. M. (2004). Social status and health in humans and other animals. *Annual Review of Anthropology, 33,* 393–418.

Savitri, A. I., et al. (2014). Ramadan fasting and newborn's birth weight in pregnant Muslim women in the Netherlands. *British Journal of Nutrition, 112,* 1503–1509.

Scholte, R., et al. (2017). Does the size of the effect of adverse events at high ages on daily-life physical functioning depend on the economic conditions around birth? *Health Economics, 26*(1), 86–103.

Shah, A. K., et al. (2012). Some consequences of having too little. *Science, 38,* 682–685.

Shu, H., et al. (2018). Prenatal phthalate exposure was associated with croup in Swedish infants. *Acta Pediatrica, 107*, 1011–1019.

Sokol, J. R., et al. (2003). Fetal alcohol spectrum disorder. *Journal of the American Medical Association, 290*(22), 2996–2999.

Statistics Sweden. (2018). Medellivslängden ökar mest för högutbildade (Average longevity increases mostly for the highly educated). *Stat. News* 2018-10-09.

Stephenson, J., et al. (2018). Before the beginning: Nutrition and lifestyle in the preconception period and its importance for future health. *The Lancet, 391*(10132), 1830–1841.

Thomas, M. S. C., & Johnson, M. H. (2008). New advances in understanding sensitive periods in brain development. *Current Directions in Psychological Science, 17*(1), 1–5.

Tobi, E. W., et al. (2018). DNA methylation as a mediator of the association between prenatal adversity and risk factors for metabolic disease in adulthood. *Science Advances, 4*(1), eaao4364 (31 January 2018).

Torssander, J., & Erikson, R. (2010). Stratification and mortality—A comparison of education, class, status, and income. *European Sociological Review, 26*(4), 465–474.

van den Berg, G., et al. (2006). Economic conditions early in life and individual mortality. *American Economic Review, 96*(1), 290–302.

van den Berg, G., et al. (2017). Mortality and the business cycle: Evidence from individual and aggregated data. *Journal of Health Economics, 56*(2017), 61–70.

Vassoler, F. M., et al. (2014). The impact of exposure to addictive drugs on future generations: Physiological and behavioral effects. *Neuropharmacology, 76*, Part B, 269–275.

Vieira, S. (2015). The health burden of pollution: The impact of prenatal exposure to air pollutants. *International Journal of Chronic Obstructive Pulmonary Disease, 10*, 1111–1121.

Vikat, A., et al. (2002). Sociodemographic differences in the occurrence of teenage pregnancies in Finland in 1987–1998: A follow up study. *Journal of Epidemiology and Community Health, 56*, 659–668.

Warrington, W., et al. (2000). Student attitudes, image and the gender gap. *British Educational Research Journal, 26*(3), 393–407.

Weisleder, A., & Fernald, A. (2013). Talking to children matters: Early language experience strengthens processing and builds vocabulary. *Psychological Science, 24*(11), 2143–2152.

Williams, B. F., et al. (1994). Fetal alcohol syndrome: Developmental characteristics and directions for further research. *Education & Treatment of Children, 17*, 86–97.

Wolbers, M. H. J. (2007). Employment insecurity at labour market entry and its impact on parental home leaving and family formation. A comparative study among recent graduates in eight European countries. *International Journal of Comparative Sociology, 48*(6), 481–507.

World Bank. (2018). *World development report 2018: Learning to realize education's promise.* Washington, DC: The World Bank.

Zhang, X., et al. (2018). The impact of exposure to air pollution on cognitive performance. *Proceedings of the National Academy of Sciences, 115*(37), 9193–9197.

Chapter 4
Modelling and Analysis of Life Cycles*

Models of life cycle changes in abilities, health and income are typically based on a capital approach. The idea is that the capital that a person carries in some sense summarises his or her previous history, which is of decisive importance to the ability to exploit new possibilities and meet new challenges.

The concept of *capital* is an old one. In the Middle Ages, it was used for the main part of a loan, excluding interest (which was called *usury* at the time).[1] The broader sense of wealth—a stock (of wealth or property, or the value of either) existing at an instant of time[2]—dates back to the seventeenth century. Originally referring to physical entities such as buildings and agricultural land, it gradually assumed the more abstract meaning of expected income flows tied to such assets.[3]

Adam Smith also used the term capital in the modern sense of human capital: "The acquisition of such talents, by the maintenance of the acquirer during his education, study, or apprenticeship, always costs a real expense, which is a capital fixed and realized, as it were, in his person."[4] The term *human capital* gained broader acceptance during the 1960s following work by Becker and Mincer.[5] It is used both in the narrow sense of abilities that are acquired by various forms of learning and formal education and, more broadly, as also including other assets such as *health* or *social networks*. In what follows, these latter assets will be dealt with separately under the headings of health capital and social capital, respectively.

---

[1] Fetter (1937). The etymological origin is the Latin *caput*, meaning *head*, metaphorically an important part, such as the capital of a country.

[2] Fisher (1904).

[3] On this development, see Commons (1924) and Horwitz (1977), both with a focus on the United States.

[4] Smith (1776), Book II, Chapter I.

[5] Becker (1964), Mincer (1970).

© The Author(s), under exclusive license to Springer Nature Switzerland AG 2022
P. Molander, *The Origins of Inequality*,
https://doi.org/10.1007/978-3-030-93189-6_4

In an attempt to widen the concept of capital further to include the total wealth of a country, *natural capital* has been added.[6] This form of capital falls outside the scope of the present discussion.

4.1 Growth Processes

Consider a collection of entities—persons, companies, cities—that receive an inflow of some sort, such as knowledge, income, immigrants, etcetera. The stock at time t is called θ_t. This inflow is subject to stochastic variations. Assume that the inflow per time unit can be modelled as independent, identically distributed stochastic variables ε_t:

$$\theta_{t+1} = \theta_t + \varepsilon_{t+1}$$

The central limit theorem of statistical theory then establishes that the size distribution will approach a normal distribution in the limit as time goes to infinity.

This result is important, but as a model of growth processes it lacks realism. In general, we expect the inflow to depend on the current state. Large cities and large capital assets will in general grow faster in absolute terms than smaller ones. Formally, if θ_t is the state at time t and ε_{t+1} is the inflow at time instant $(t + 1)$, we can write

$$\theta_{t+1} = \theta_t + f(\theta_t, \varepsilon_{t+1}, t) \tag{4.1}$$

The behaviour of this process will depend critically on what is assumed about the dependence of f on θ. For simplicity, assume that there is no explicit dependence on t. A natural and commonly used assumption is that growth is proportional to the current state, that is,

$$f(\theta_t, \varepsilon_{t+1}, t) = k \cdot \theta_t \cdot \varepsilon_{t+1}$$

for some $k > 0$. This yields

$$\theta_{t+1} = \theta_t + k \cdot \theta_t \cdot \varepsilon_{t+1} = (1 + k\varepsilon_{t+1}) \cdot \theta_t = \theta_0 \cdot \prod_{i=1}^{t+1} (1 + k\varepsilon_i)$$

Taking logarithms, we get

---

[6] See Hamilton and Hepburn (2017) for an overview of this research.

$$\log (\theta_{t+1}) = \log (\theta_0) + \sum_{i=1}^{t+1} \log (1 + k\varepsilon_i)$$

By the central limit theorem, it can be concluded that the logarithm of θ_t in the limit will be normally distributed. In other words, assuming that growth is proportional to the current state leads to a *lognormal distribution* of the size.[7]

The assumption that growth is proportional to the current state, or, as otherwise expressed, that the growth rate is independent of the size, is referred to as *Gibrat's law*.[8] Gibrat illustrated his analysis with examples on firm sizes, incomes, city populations and other areas. One can plausibly imagine that growth rates are both lower and higher than proportional to θ. If there is a limit to the capacity to receive the inflow, because of bottlenecks or saturation, growth will be slower than proportional. In the case of cities, by contrast, one can imagine the same natural net population growth for all cities, but if the inflow of migrants into large cities is stronger because of a more attractive labour market, growth may be faster than proportional. Firm size distribution can be affected by returns to scale. Similarly, large capital owners can afford to buy information beyond what is generally available in financial markets and in this way achieve a higher-than-normal growth rate. In such cases, the resulting distribution will have a fatter tail than the lognormal distribution.[9]

Power Laws

The concept of *power laws* is relevant in this context. A distribution is said to obey a power law if the frequency function $f(x)$ satisfies

$$f(x) \sim c \cdot x^{-\alpha}$$

for some constants c and $\alpha > 0$. The distribution is called *scale-free*.

Power laws appear in a large number of natural and social systems—earthquakes, solar flares, frequency of word use, networks, city populations, intensity of wars and others. The exponent α typically lies in the interval [2,3]. If $\alpha \leq 2$, the distribution will not have a finite expected value.

If X is lognormally distributed with mean μ and standard deviation σ, the frequency function $f(x)$ will satisfy

---

[7] For a comprehensive treatment of the lognormal distribution, see Crow and Shimizu (1988).

[8] Gibrat (1930). Other important early analyses of skew distributions are Pareto (1897), Yule (1925) and Simon (1955).

[9] For surveys on the genesis of lognormal distributions and power laws in physics, biology, economics and networks, see Newman (2005), Scheffer et al. (2017), Gabaix (2016) and Mitzenmacher (2004), respectively.

$$f(x) = \frac{1}{\sqrt{2\pi}\sigma x} \cdot e^{-(\ln(x)-\mu)^2/2\sigma^2}$$

Taking logarithms,

$$\ln f(x) = -\left(1 - \frac{\mu}{\sigma^2}\right)\ln x - \frac{(\ln x)^2}{2\sigma^2} - \ln \sqrt{2\pi}\sigma - \mu^2/2\sigma^2$$

If σ is large, it follows that the lognormal distribution will resemble a power law over a large interval. For real-life distributions, a common solution is to approximate the distribution observed with a combination of a lognormal and a genuine power law distribution, or with a combination of power law distributions.

Consequences for Inequality
In the perspective of inequality, the important conclusion concerning processes following Eq. (4.1) is that the variance of the process will increase with time, and that the distribution will be increasingly skewed to the right. This holds independently of the original distribution. For the special case of proportional growth (Gibrat's law), logarithmic variance will increase linearly with time.

More precisely, if the average of the underlying distribution is μ and the standard deviation is σ, the lognormal distribution resulting asymptotically from Eq. (4.1) will be characterised by the following basic parameters:

- *average*: $\exp.(t\mu + t\sigma^2/2)$
- *variance*: $\exp.(2t\mu + t\sigma^2)\cdot[\exp(t\sigma^2) - 1]$.

Different measures of inequality will take the following form:

- *coefficient of variation*: $[\exp(t\sigma^2) - 1]^{\frac{1}{2}}$
- *skewness*: $[\exp(t\sigma^2) + 2]\,[\exp(t\sigma^2) - 1]^{\frac{1}{2}}$
- *Gini coefficient*: $2\Phi(\sigma\sqrt{t/2}) - 1$, where $\Phi(\cdot)$ is the distribution function of the normal distribution $N(0,1)$.

In summary, inequality grows without limit, as time tends to infinity, whatever measure of inequality is chosen.

Intergenerational Transmission
Equation (4.1) refers to an entity of infinite longevity, which may be relevant for cities or legal persons such as corporations or trust funds. For human beings, transmission between generations becomes important. Borgerhoff Mulder et al. (2009) use a difference equation to model the intergenerational transmission of wealth. If w_i is the logarithmic wealth of household i, the combined effect of parental and non-parental influences on a household's wealth can be expressed as $\beta w_{ip} + (1 - \beta)w_{av}$, where w_{ip} is the wealth level of household i's parents, w_{av} is the population-average wealth level (normalised to be constant), and β $(0 \leq \beta < 1)$ is the intergenerational transmission coefficient. In each transmission between generations, there is a disturbance term, ε, assumed to be independent of the wealth of

previous generations, with zero mean and variance σ_ε^2. For the variance of household wealth, a straightforward calculation yields

$$var(w_i) = \sigma_\varepsilon^2 / (1 - \beta^2)$$

A high transmission factor is consequently equivalent to a large variation in assets and a high level of inequality. In societies where material assets, physical or financial, are important, there will be a stronger drive towards high inequality. Not only the character of the dominant asset but also inheritance legislation is important—whether women have the right to inherit, the existence of entailed estate, and so on. Most of human history can be characterised by an increasing β, which adds to the basic drive towards growing inequality that is inherent in Gibrat's principle. The major exception, in recent times, is the increasing effect of human capital, for which the transmission coefficient is lower than for material assets.

The idea that transmission takes place only between parents and children leads to low transmission effects over longer periods, even if the transmission coefficient is high. Becker and Tomes assert that "almost all earnings advantages and disadvantages of ancestors are wiped out in three generations".[10] This view has proved to be incorrect when studies have been extended to cover *multigenerational transmission*.[11] More precisely, consider some status variable $y_{i,t}$ of a child in household i at time t, which may refer, for example, to income or education. If transmission is proportional with coefficient β_{-1} and ε is an error term, one gets

$$y_{i,t} = \alpha + \beta_{-1} y_{i,t-1} + \varepsilon_{i,t}$$

For correlations between more distant generations, assuming constancy, one would get $\beta_{-m} = (\beta_{-1})^m$, confirming the statement by Becker and Tomes. Persistency has in reality turned out to be higher, as shown in a number of studies. Clark and Cummins use surnames as indicators and claim a high level of persistency.[12] Lower levels, though still higher than in the proportional model, have been found in later studies. This higher level of persistency can be captured using a *latent factor*[13]:

$$\begin{cases} y_{i,t} = \rho e_{i,t} + \delta_{i,t} \\ e_{i,t} = \lambda e_{i,t-1} + \varepsilon_{i,t} \end{cases}$$

Here, λ is a coefficient of inheritance and ρ a coefficient of transfer. If the link between outcomes and latent factor is perfect ($\rho = 1$), this is identical to the proportional model, but if it is imperfect ($\rho < 1$), $\beta_{-m} > (\beta_{-1})^m$, and persistency

---

[10] Becker and Tomes (1986).

[11] Lindahl et al. (2015) and the following notes.

[12] Clark and Cummins (2015) (UK data).

[13] Braun and Stuhler (2018).

will be higher. Previous results on country differences have been confirmed, for instance that persistency is higher in the United States than in Sweden.[14]

The importance of the extended family, the dynasty, for the persistence in human capital inequality across generations is verified in a study by Adermon et al.[15] In addition to parents, grandparents and great grandparents, parents' siblings and cousins, as well as their spouses and the spouses' siblings are included. The persistence across generations estimated by the extended model is about twice as large as traditional parent-child estimates.

Assortative Matching

What happens in the transition between two generations is not only a transfer of assets of various kinds but also the formation of new families. Depending on the matching between partners, inequality may be affected more or less strongly. If new couples are formed at random, matching has an equalising effect. The *degree of assortativity* can thus be expected to affect the general level of inequality in society.

Fernández and Rogerson used a simple model to analyse the effects of sorting on inequality and found significant effects when applying it to data from the United States.[16] Several factors intervene. A negative correlation between education and fertility, a decreasing marginal effect of parental education on that of children and a wage formation process that is sensitive to the relative supply of skilled labour all contribute to strengthening the link between sorting and inequality. In this model, both fertility and the matching process are exogenously determined. When low family income is coupled to high fertility and greater wage differentials intensify matching with high-income individuals, conclusions are reinforced.

Greenwood et al. confirm a picture of increased assortative matching in the United States.[17] The high level of female labour-force participation among married women is important for this outcome.

In a broader study of educational sorting, Eika et al. confirm the picture of assortative matching at all levels of education in the countries studied.[18] Time trends vary by the level of education; among college graduates, assortative mating has been declining over time, whereas the opposite holds among persons with lower levels of education. At a given point in time, educational assortative mating accounts for a significant part of inequality in household income, but over time, trends in household income inequality are largely unaffected, because of the balancing effects in matching patterns at different levels of education.

Collado et al. combined a model of latent effects with assortative matching and found a much higher degree of sorting than in previous studies.[19]

[14] Vosters and Nybom (2017).

[15] Adermon et al. (2021) (Swedish data).

[16] Fernández and Rogerson (2001).

[17] Greenwood et al. (2014).

[18] Eika et al. (2019). Data from the U.S., Denmark, Germany, the UK, and Norway.

[19] Collado et al. (2020) (Swedish data).

Notes on Statistical Distributions

The normal distribution is the dominant model for characterising quantitative data from populations. There may be many reasons why actual data should deviate from normal distributions, however. From the above discussion, it is clear that whenever multiplicative rather than additive processes are involved, a lognormal rather than a normal distribution of the outcomes should be expected. Practical estimates could gain from this simple observation.[20]

Estimating power law distributions gives rise to specific problems. Analyses often start from a visual inspection of a log-log diagram, and there is a risk of erroneous conclusions, given that all samples are finite. Straightforward linear regression on logarithmic data may yield biased estimates, but in some situations there exist simple alternatives that evade this problem.[21]

Empirical estimates of human capital via raw data from aptitude tests deviate from normal distributions. The reason is that tests tend to be too easy and that a large part of the scores will therefore cluster at the upper end of the distribution. A detailed analysis of aptitude tests will confirm that the upper tail is thick as predicted, although this may not show up in the skewness measure.[22] The fact that aptitude tests often neglect the important non-cognitive component of human capital is a different problem.[23]

4.2   Human Capital

The interest in human capital was sparked by Solow's observation that only a limited part of economic growth could be explained by increasing input of labour (number of working hours) and an increasing capital stock.[24] A significant part of the unexplained growth is nowadays ascribed to human capital.

The amount of human capital at the individual or aggregate level in a country can be measured or estimated using different approaches. One approach measures human capital by looking at the total stream of past investments from the individual, the family, employers and governments. These costs include both monetary outlays and non-market expenditures, such as the time devoted to education. A second

---

[20] Limpert and Stahel (2011).

[21] Newman (2005), § II.

[22] This was observed already by Lord (1955) and Cook (1959), and later confirmed by Micceri (1989). The title of Herrnstein's and Murray's (1994) contested book, *The Bell Curve*, is a misnomer. The distribution of data from the Armed Forces Qualifying Test, which is the basis of their argument, is truncated to the right, strongly skewed to the left, and conforms to a normal distribution only after a heavy-handed analytical transformation (see Fischer et al. (1996), Fig. 2.1). A thorough analysis of the statistical problems associated with these tests can be found in Ho and Yu (2015).

[23] Cunha et al. (2006).

[24] Solow (1957).

approach measures human capital through literacy and aptitude tests and years of schooling. A problem with this approach is that it works well at the individual level but lacks a common metric and therefore cannot be aggregated. Comparisons of the total stock of human capital across countries and over time become difficult.

A third approach measures human capital at the output side by looking at the stream of future earnings that human capital investment generates. This is the method used by the OECD in international comparisons.[25]

Estimates of human capital assets at the national level indicate that their relative importance grows with economic development.[26] In advanced OECD countries, the ratio of the value of human capital to the total value of physical capital typically lies between 4 and 5.[27]

When introducing human capital as a production factor, it is important to observe a number of important differences between human and physical capital. Firstly, at a given instant, the total volume of physical capital is given. If a certain amount of physical capital is transferred between two individuals, an increase in the possession of one party is always matched by a decrease in the possession of the other. Information or knowledge, by contrast, does not come in fixed amounts. Sharing knowledge with someone will not necessarily imply a loss; in fact, teaching someone else a task can improve one's own level of understanding. Certain types of knowledge can lose value if shared with a large number of people, but that is another matter.

Secondly, physical capital loses value when used, due to wear and tear, whereas human capital, by contrast, improves. "Use it or lose it" is a common adage that is tied to the body and physical abilities but is equally relevant for the brain and mental activities.[28]

With these reservations in mind, it may nonetheless be reasonable to try to model human capital via a variable that summarises relevant abilities, cognitive as well as non-cognitive, and the effects of prehistory. Learning is a highly complex process involving various parts of the human brain and the environment, which might be a classroom but in fact may be any situation where human cognition is involved. As a consequence, the analysis of learning must involve a whole range of disciplines, from neurophysiology via cognitive psychology to pedagogics. Analysing the brain as a network, and consequently the learning brain as a growing network, is an approach that has made important progress in recent decades but is still at an early stage due to the complexity of the brain.[29]

Neuroimaging has made it possible to see the physiological consequences of both learning and adverse events. When a child learns to read and to count, it develops not only those capabilities but also more general ones. A literate person, when facing a

---

[25]For a discussion of methodological problems, see Liu (2011).

[26]Hamilton et al. (2017).

[27]Liu (2011).

[28]Blakemore and Frith (2005).

[29]See Sporns (2011) and Bassett and Mattar (2017).

given problem, activates a larger part of the brain than an illiterate one.[30] This process continues into adulthood. Conversely, when adverse events such as accidents or illness impair a part of the brain, other parts may to some extent compensate for the loss. *Generality* and *plasticity* are thus two important characteristics of brain functioning.[31]

Basic Structure

As a preparation for the model building, a number of stylised facts can be summarised:

- Skills comprise both cognitive and noncognitive factors. Both are important in a child's development into adolescence and adult life. They may be mutually supportive, but they develop along different tracks.
- Gaps between individuals are visible at an early age and tend to grow with age. Genes play a role, but it has traditionally been exaggerated.
- Development after conception is not uniform but goes through sensitive and sometimes critical periods. Early life is generally more important than later periods.
- During school years, the development of skills dispersion depends on a number of factors in the design of the educational system.
- Family background is important both as a source of development of skills and as a resource in the case of adverse events.[32]
- For individuals in the lower part of the distribution of human capital, there may be bifurcation points at which the fundamental attitude to learning switches from positive to negative, with far-reaching consequences for later phases of individual development.[33]

When a group of persons are exposed to a flow of information, for instance pupils in a classroom, they will not absorb this information at equal rates. New information or knowledge requires a point where it can be anchored, and the likelihood of finding such a point increases with the amount of human capital available. In the first approximation, the likelihood of hitting the target can be assumed to be proportional to the size of the target, that is, to the existing human capital. This is reminiscent of the growth processes described in the previous section, and the corresponding mathematical tools and results become relevant.

In a series of publications, Heckman and associated researchers have argued for a model of human capital formation that is built on this recursive approach.[34] A basic form is

---

[30] See Castro-Caldas et al. (1998) and Looi et al. (2016) concerning the effects of literacy and numeracy, respectively.

[31] Davison et al. (2015).

[32] Fredriksson et al. (2016), Almond et al. (2018).

[33] Warrington et al. (2000), Gustafsson et al. (2010).

[34] Cunha et al. (2006), Heckman (2008), Cunha and Heckman (2009).

$$\theta_{t+1} = f(h, \theta_t, I_t, t) \tag{4.2}$$

Here, h represents the ensemble of factors associated with parental abilities (genes, income etc.), and I_t is the investment at time t in the child's human capital. Compared to Eq. (4.1), the explicit family background and the investment are new, whereas the stochastic element ε_t is left out.

An important feature of this model is that childhood consists of a number of stages, and that actions taken in different stages are interdependent. Further, in line with empirical research, the function f is assumed to satisfy the following conditions:

$$\text{Self-productivity}: \qquad \frac{\partial f(h, \theta_t, I_t, t)}{\partial \theta_t} > 0$$

$$\text{Dynamic complementarity}: \qquad \frac{\partial^2 f(h, \theta_t, I_t, t)}{\partial \theta_t \partial I_t} > 0$$

The first condition is self-evident; it is hard to imagine a process by which human capital that is larger at time t would not also be higher at time $(t + 1)$. The second condition is non-trivial, implying that the larger the stock is, the more productive are investments.

The basic model can be refined in various ways. Human capital can be split into cognitive and noncognitive components, justified by the fact that they respond differently to investments and also appear to have different effects in the labour market.[35] Differences in personality may be as least as important as differences in cognitive skills,[36] and there may also be complementarity between cognitive and non-cognitive abilities. Self-confidence or perseverance may improve learning, and knowledge strengthens self-confidence. In mathematical terms, if θ_i and θ_j are components of the $\vec{\theta}$ vector,

$$\frac{\partial^2 f\left(h, \vec{\theta}_t, I_t, t\right)}{\partial \theta_{i,t} \partial \theta_{j,t}} > 0 \tag{4.3}$$

Equation (4.2) is very general and must be specified in order to be practically useful. One should not expect the same specification to be appropriate during the whole life cycle. As a minimum, three models seem necessary: one for childhood and adolescence, one for adulthood and one for old age.

Childhood and Adolescence
During *childhood and adolescence*, investment decisions are made by the parents and the public sector on behalf of the child. The investment term should be split into private investments made by the family and public investments in the system of

---

[35] See Kautz et al. (2014) for a survey.
[36] Borghans et al. (2011).

education (I_t^{fam} and I_t^{publ}, respectively). These two have different characteristics and different effects on outcomes with respect to inequality. The term *investment* may exaggerate the degree of consciousness in decisions made, or not made, during the childhood and adolescence periods. A lack of information or knowledge limits the spectrum of alternatives considered by many parents, and decisions may be guided by habit and tradition rather than by serious consideration of the pros and cons of the alternatives actually available. Uncertainty is present in all households irrespective of socioeconomic status, given that parents have limited information about schools and how to interpret it. They may also maintain incorrect perceptions about the abilities of their children. Peer effects may be difficult to predict.

The risk of underinvestment must be acknowledged. Education has non-market private values, which are difficult to estimate, and collective values, which would be neglected even if they were correctly estimated.[37]

In order to avoid double-counting, the factor h should be reserved for that part of the family background that refers to prenatal and perinatal conditions. Further, in order to acknowledge the boundedness of rationality and the uncertainty of all investment decisions, it is preferable to keep the stochastic component from Eq. (4.1).

In summary, an equation for the first phase would be

$$\vec{\theta}_{t+1} = f\left(h, \vec{\theta}_t, I_t^{fam}, I_t^{publ}, \varepsilon_{t+1}, t\right) \qquad (4.4)$$

Family investments can be expected to increase inequality whatever level of inequality in the distribution of human capital exists at birth. Whether public investments in the form of schooling will reduce or increase inequality depends on both educational policies at large and teaching methods. The spread of human capital distribution will be reduced as school raises the lowest level for the pupils with the least favourable background conditions. If teaching on top of that favours strongly adaptive pedagogical methods, inequality will be further reduced. If focus is instead on the middle of the distribution, the least favoured pupils will lose in relative terms, and middle and upper ends of the distribution will gain. If focus is even higher up the distribution, or if student-centred or enquiry-based pedagogical methods are applied, inequality will be further increased. Whether overall inequality will increase will consequently depend both on teaching methods and on the measure of inequality used.

What is not captured by Eq. (4.4) is the type of qualitative changes or bifurcations that result in negative attitudes to learning. For this type of phenomenon, different mathematical models must be used. There are examples of such models that have been applied to learning processes. The main focus appears to be on limited aspects of learning, however, such as sensorimotor processes.[38] In general, this seems to be an underexploited approach.

---

[37] McMahon (2009, 2018).

[38] Kostrubiec et al. (2012) is an illustrative example.

Adulthood

In the *adult phase*, the individual first faces the choice between going directly to the labour market or going to higher education and then to the labour market. In either case, choices at this level can be considered autonomous, so investments should be modelled as functions of the current state of human capital: $I_t = g(\theta)$ for some function $g(\cdot)$. Investments can be assumed to increase with θ, and a proportional relationship is reasonable. If it is further assumed that the stochastic element can be described in multiplicative terms, we get

$$\theta_{t+1} - \theta_t = k \cdot \theta_t \cdot \varepsilon_{t+1}, \; k > 0 \tag{4.5}$$

This equation is identical to Eq. (4.1), generating a lognormal distribution of human capital with increasing variance. An increasing spread in earnings or consumption within cohorts has been confirmed in several studies, in conformity with the above equation.[39]

More developed models have to take account of job mobility and seniority, as well as various shocks that may occur during working life.[40] The experience profile of wages varies between professions. In professions where the accumulation of empirical knowledge is important, such as among lawyers, human capital can be assumed to grow throughout the entire professional career. At the other extreme, professional athletes often retire in their 30s, and so either have to switch to a different career or accumulate enough capital during their active period to be economically independent. Job changes and unemployment are important events that may have long-lasting effects on earnings development. Human capital accumulation nonetheless accounts for most of the growth of earnings.

Old Age

In the long run, human capital will necessarily start to decrease because of *aging*. Transition to the retreat phase may be more or less protracted and occurs at different ages in different professions.

During the old-age phase, human capital will depreciate, and the standard model of decay is appropriate:

$$\theta_{t+1} - \theta_t = -\delta \cdot \theta_t \cdot \varepsilon_{t+1}, \; \delta > 0 \tag{4.6}$$

After retirement, there is consequently a tendency towards convergence—how strong depends on the environment and, as a consequence, indirectly on the design of pension systems, savings patterns, etcetera.

---

[39] Creedy (1985), Deaton and Paxson (1994), O'Rand (1996), Crystal et al. (2017).
[40] See Altonji et al. (2013) for an example.

4.3 Health Capital

Health, like skills, is an important dimension of human well-being. Like human capital, health capital at the individual or aggregate level can be measured in different ways. One approach uses the discounted value of future life-years. Adjustments for quality can be made for health status using quality-adjusted life years (QALYs) or disease-adjusted life years (DALYs). Using this method, it is possible to identify longitudinal changes in the distribution of health capital. As noted previously (Sect. 3.5), the distribution of QALYs in Sweden has become more unequal during the last two decades of the twentieth century.[41]

A different approach is to integrate the health factor into an economic analysis and to isolate the contribution from good health to income at the individual level, or to GDP or total assets at the aggregate level. When the economic importance of health is estimated in this way, the conclusion is that it is important, although less so than skills. Differences in health account for 10 to 12 per cent of GDP levels both in longitudinal change and in cross-country variations.[42] Locally and regionally, health may be more important. Some estimates of the gain from eradicating malaria in heavily infested regions indicate a much higher payoff.[43]

A number of stylised facts should be accounted for:

- Health is strongly associated with socioeconomic status, and education seems to be the most important component of the bundle of status-defining variables, followed by occupation and health-related behavioural patterns.
- A significant proportion of health variations related to socioeconomic status can be related to family background. Health inequalities between low- and high-status groups are already measurable at birth, and they increase over the life cycle.[44]
- In old age, there is some tendency for these inequalities to narrow, although the picture is somewhat diverse, and some of the effect may be due to culling.[45]

These facts taken together suggest a capital model for health similar to the one used for skills.[46] A commonly used model is the one suggested by Grossman,[47] which in the notation used here runs

$$\theta_{t+1} = (1 - \alpha_t) \cdot \theta_t + f(M_t, T_t, E)$$

---

[41] Burström et al. (2005).

[42] Weil (2014).

[43] Gallup and Sachs (2001). Other estimates have been more modest.

[44] Case et al. (2002).

[45] Case and Deaton (2005).

[46] Early suggestions in this direction were made by Schultz (1962) and Becker (1964); see the survey by Becker (2007).

[47] Grossman (1972, 2000).

Here, α_t is a natural rate of deterioration of health, M_t is medical care and T_t time spent on medical care, and E is education. The focus here is on adulthood and old age, and the path followed is controlled by investments in health, besides the natural aging process. This model has met with some criticism; it focuses on adulthood and old age rather than the beginning of life, and investments in health care are considered by many observers less important than other factors. Although the amendments made take care of some of the objections, the main problems remain.[48]

A different approach, given the obvious parallels between health capital and skill-based capital, is to extend the previous model for human capital by including health as a component of the state of human capital, θ_t^H. Further, if skills are separated into cognitive and non-cognitive skills, we get a vector, $\overrightarrow{\theta_t} = (\theta_t^C, \theta_t^N, \theta_t^H)$, for which the following equation can be assumed to hold[49]:

$$\overrightarrow{\theta}_{t+1} = \overrightarrow{\theta}_t + f\left(h, \overrightarrow{\theta}_t, I_t, \varepsilon_{t+1}, t\right) \tag{4.7}$$

As before, h summarises the background, which, together with the current state of cognitive skills, non-cognitive skills (personality), health stock and effort (investment) determines outcomes. The model can be refined by breaking down the three main components into more detailed descriptions. For instance, efforts have been made to link characteristics from economic psychology such as patience and perseverance to traditional personality traits, the so-called "Big Five".[50] A variety of outcomes have been analysed, such as wages, crime, health and health-related behaviours.

Complementarity according to Inequality (4.3) has been confirmed.[51] Health at the age of 10 has a statistically significant direct effect on adult outcomes at the age of 30. Importantly, there is also evidence of complementarity between the different components of the capital vector, at least in the weaker sense that an increase in one component has nonnegative effects on other components. High non-cognitive skills increase the likelihood of healthy behaviours; their role is strong, and if they are neglected, the importance of cognitive factors will be overestimated. Personality traits explain more than half of the observed differences in poor health and obesity via selection, but there is also a verified causal effect of education on health and health-related behaviour, for instance smoking.

It is important to recognise the heterogeneity of effects, for instance between men and women. By way of example, early health endowments are associated with schooling for women but not for men.

---

[48] Galama and van Kippersluis (2013).

[49] Heckman (2012).

[50] Almlund et al. (2011). More recent research has led some psychologists to suggest six rather than five traits; see Ashton et al. (2014).

[51] Conti and Heckman (2010), Conti et al. (2010) (UK), Heckman et al. (2018) (USA) concerning the statements that follow.

4.4 Social Capital

If the concepts of human capital with respect to skills and abilities and to health are somewhat vague and correspondingly difficult to measure in an uncontroversial manner, this is all the more true for social capital.[52] The modern use of the concept can be dated to the late 1970s, introduced by Loury, Bourdieu and Coleman.[53] Definitions vary, from relatively well delineated to very wide. According to Loury, social capital is the set of resources that are inherent in family relations and in community social organisation and that are useful for the cognitive and social development of a child or young person. For Bourdieu, social capital is a sustainable network of relations that are useful to its members. Putnam's definition is extensive: "connections among individuals—social networks and the norms of reciprocity and trustworthiness that arise from them".[54] Others see the triplet of trustworthiness, networks and (more or less formalised) institutions as the basis, while trust is the link between this triad and collective action.[55]

A somewhat more precise network-based definition was developed by Burt, differentiating with respect to the complexity of the network and whether its structure is relational or positional.[56] The development of network theory has paved the way for even more precise and mathematically based definitions.[57]

Social capital can be measured on different levels, from the family and small groups to whole countries. In general, there is no necessary connection between social capital on different levels. High local trust may well combine with mistrust at the aggregate level.

It is common to classify social capital as a collective good, but there are nonetheless important individual variations in the access to this collective good.[58] This is obvious for the network-based definitions but holds true also for more abstract versions that identify social capital with trust.

The choice of indicators for the measurement of social capital obviously depends on how the concept is interpreted. Social capital is measured basically in two ways, either via visible indicators such as frequency of memberships in organisations or via surveys of the level of generalised trust in human beings.[59] The first group focuses

---

[52] For a survey of the literature, see Lyon et al. (2015).

[53] Loury (1977), Bourdieu (1980), Coleman (1990), Chap. 12. For an overview of the different uses and the development, see Häuberer (2011).

[54] Putnam (2000).

[55] Ostrom and Ahn (2009).

[56] Burt (1982).

[57] We will return to the formal theory of networks in Chaps. 5 and 6.

[58] Poortinga (2002).

[59] The standard indicator is the frequency of persons supporting the idea that "[m]ost people can be trusted". For a general discussion of measurement problems connected with social capital, see Scrivens and Smith (2013).

on local communities, the second on more general and abstract social entities, so these indicators may point in different directions.

A few stylised facts for this form of capital are:

- Social capital tends to correlate positively with human capital and health capital.[60] Persons with higher education and high incomes tend to be well supplied also with social capital. Some qualifications to this statement are necessary, though. Individuals with a high level of trust and social engagement more often report good health than individuals with lower levels in countries with high levels of social capital, but the reverse relationship holds in countries with low levels of social capital.
- Social capital resembles the other forms of capital in the sense that the relationship between social capital and age first increases and then decreases in old age.
- A specific characteristic of social capital is the dependence on space. Social capital at local and intermediate levels falls with physical distance.
- There is a strong correlation between equality and social capital defined as trust.[61]

In general terms, the variation of social capital across social groups and over the life cycle follows a pattern and a dynamic similar to those of human or health capital. The more pronounced collective and abstract character of social capital makes it less accessible to analysis, and there is relatively weak consensus on the determinants and the effects of social capital.[62]

The Significance of Social Capital
Even though the definition and measurement of social capital is fraught with difficulties and consensus does not prevail, it may be interesting to get a feeling for its order of magnitude at the macro level. Hamilton et al. have attempted such an estimate based on the measurements of subjective well-being that have been carried out at the global level.[63] In this approach, subjective well-being is modelled as a function of different capital components—human, social, etcetera—which makes it possible to compute the exchange value between these different components. Social capital is both a direct source of well-being and a productive input.

The measurements of social capital, based on the standard binary question of trust in fellow human beings, illustrate large differences between countries when it comes to the value of social capital, both in absolute and relative terms.[64] In general, the aggregate value of social capital increases both absolutely and relatively with the general level of development of a country, but trust levels can also differ significantly between countries at comparable levels of GDP. For instance, the difference

---

[60]For a summary, se Glaeser et al. (2002).

[61]Uslaner and Brown (2005), Bjørnskov (2007), Jordahl (2009), Rothstein (2011), Barone and Mocetti (2016).

[62]For a critical review of a few benchmark analyses, see Durlauf (2002). See also Uslaner (2018) and Chaps. 5 and 6.

[63]Hamilton et al. (2017).

[64]Ibid., Table 12.1 and Fig. 12.1.

in trust levels between Sweden and Italy corresponds to an increase in Italian GDP per capita of 20 per cent, if Italy were to achieve the same level of trust as Sweden currently exhibits.[65]

Intergenerational Transmission of Social Capital

The transmission between generations of social capital obviously depends on what aspect of social capital is in focus. As for networks, these are mainly private assets that parents attempt to carry over to their children via choice of schools, leisure time activities and so on. The general trust level is a more pronounced collective asset, the transmission of which depends on factors beyond the reach of the single household. At the same time, basic values are instilled in children as part of their education, so parents also play an important role in this respect.

Levels of general trust measured among migrants and their descendants exhibit both stability and change. Several measurements of trust level in American states have been made[66]; the single-most important migration-related factor is the ratio of immigrants from the Nordic countries.[67] States with a high historic ratio of Scandinavian and German immigrants display higher social mobility, both among whites and African American citizens.[68] Some studies have focused on change and adaptation to the patterns of the new country of residence. A broad study comprising 130 countries indicates that the recipient country is more important than the country of origin for the level of trust.[69] The relationship is asymmetric: migration from low-trust countries to high-trust countries exhibits a lower level of adaptation than migration in the other direction. Trust in the administration and the political system can be relatively rapidly established, however.[70]

4.5   Inequality and Social Mobility

Intergenerational transmission follows a number of different channels—biological, behavioural, financial and others. Of particular interest to the complex of inequality-related issues are the links that have been identified between levels of inequality and the strength of intergenerational transfers. High levels of inequality are correlated with low mobility. Because of its relevance to policy making, this stable relationship has given rise to a large literature on mechanisms that might explain the correlation in question.

---

[65] Hamilton et al. (2017). This figure is based on an exchange ratio between trust and income of 0.5, which is conservative. Helliwell et al. (2018) (Table 18.1) estimate an interval of between 0.7 and 0.9.

[66] Rice and Feldman (1997), Uslaner (2008).

[67] Putnam (2001).

[68] Berger and Engzell (2019).

[69] Helliwell et al. (2016).

[70] Ibid.

The negative correlation between inequality and mobility—what in colloquial language has been called the *Great Gatsby curve*[71]—is a cross-sectional relationship between some measure of inequality, in most cases the Gini coefficient, and intergenerational persistence of some variable of interest, most often income. This curve has been proved to exist in a number of countries and regions.[72] Correlation is different from causation, however, and causality could in principle go in both directions. High mobility indicates relatively equal opportunities, and given that skills tend to be more equally distributed than incomes, the important link may be the one from mobility to equality. Conversely, stark inequalities in a society may cause educational opportunities to be highly dependent on parents' incomes and social status, thereby leading to high intergenerational persistence of both incomes and other indicators of well-being.

A dynamic theory is necessary in order to describe in more detail what mechanisms may be responsible for the intergenerational persistence observed. Economic analyses quite naturally tend to be based on economic determinants. Becker et al. focus on *differences in wealth* among parents and show that even when capital markets are perfect and children have identical innate abilities, wealthy parents will invest more in their children than poor ones, and inequalities will persist particularly at the top of the distribution.[73] In a related approach, Lee and Seshadri have identified borrowing constraints as a driver of intergenerational persistence.[74]

Durlauf and Seshadri focus on neighbourhoods and identify *segregation* as a main driver behind the Gatsby curve.[75] The model that maps the relationship between incomes (Y_i) of parents (p) and children (c) belonging to household i is the following:

$$Y_{ic} = \alpha + \beta\left(Y_{ip}\right) \cdot Y_{ip} + \varepsilon_{ic}$$

for some constants α and β and household i. The important element here is that the coefficient β depends on Y, which creates a nonlinear relationship between parents' and children's incomes. This nonlinearity is essential. Another important aspect of the analysis is that inequality is not appropriately described by broad measures such as the Gini coefficient. Segregation is not generated uniformly but predominantly at the top and at the bottom of the distribution.

Under these conditions, adult incomes—as usual—can be predicted from the human capital that they have accumulated earlier in life. This accumulation of human capital is spatially determined in the sense that different income groups tend to cluster in neighbourhoods and local responsibility for the budget for education will differ as a result of differences in the tax bases. There are returns to scale in

---

[71] Krueger (2012).

[72] See Sect. 3.1.

[73] Becker et al. (2018).

[74] Lee and Seshadri (2019).

[75] Durlauf and Seshadri (2017).

schooling, which will attract parents to large communities. On the other hand, smaller but richer communities are attractive because of their stronger tax base. The resulting pattern of segregation will be determined by the interaction of these countervailing forces. Peer effects and imitation patterns will add to the financial differences between neighbourhoods.

Greater cross-sectional inequality will strengthen segregation patterns and generate larger disparities in human capital investment. This is one explanation for the causal link between cross-sectional inequality and mobility.

Both the above models have been estimated from data from the United States. In countries where higher education is free of charge and local finance is affected by more or less ambitious systems of equalisation, the determining limitations in these models will be less strongly felt, which can to some extent explain the weaker Gatsby effect for instance in Sweden.[76] Segregation is nonetheless a problem, and other, soft mechanisms of social reproduction may carry relatively more weight. Peer effects are important, and among immigrants, cultural identity seems to play a larger role than socioeconomic status.[77]

4.6 Summary

- The different forms of capital used in life cycle analysis—human, health and social capital—have similar structures and display similar patterns across the life cycle. Gaps between individuals open up early in life and tend to grow over time. There is some tendency towards convergence in old age.
- The co-variance between different components of human capital observed in empirical investigations is reflected in the dynamic complementarity of these components and investments made therein.
- The generic model of multiplicative growth covers a large part of the empirical data on life cycles. This yields a lognormal distribution for cohorts, the variance of which increases over time.
- Family background is important for all three capital forms. Intergenerational transmission is strong, though more so for financial assets and incomes than for abilities, health and social capital. Transmission across several generations is stronger than standard single-generation transmission models would suggest.
- The link from inequality to mobility, the so-called "Gatsby curve", can be given a causal explanation.

---

[76]Brandén (2019).

[77]Åslund et al. (2011, 2015).

Bibliography

Adermon, A., et al. (2021). Dynastic human capital, inequality and intergenerational mobility. *American Economic Review, 111*(5), 1523–1548.

Almlund, M., et al. (2011). Personality psychology and economics. Ch. 1 in E.A. Hanushek et al. (Eds.), *Handbook of the economics of education*, Vol. 4, 1–181. Amsterdam: Elsevier.

Almond, D., et al. (2018). Childhood circumstances and adult outcomes: Act II. *Journal of Economic Literature, 56*(4), 1360–1446.

Altonji, J., et al. (2013). Modeling earnings dynamics. *Econometrica, 81*(4), 1395–1454.

Ashton, M. C., et al. (2014). The HEXACO honesty-humility, agreeableness, and emotionality factors: A review of research and theory. *Personality and Social Psychology Review, 18*(2), 139–152.

Åslund, O., et al. (2011). Peers, neighborhoods and immigrant student achievement: Evidence from a placement policy. *American Economic Journal: Applied Economics, 3*(2), 67–95.

Åslund, O., et al. (2015). Childhood and family experiences and the social integration of young migrants. *Labour Economics, 35*, 135–144.

Barone, G., & Mocetti, S. (2016). Inequality and trust: New evidence from panel data. *Economic Inquiry, 54*(2), 794–809.

Bassett, D., & Mattar, M. G. (2017). A network neuroscience of human learning: Potential to inform quantitative theories of brain and behavior. *Trends in Cognitive Science, 21*(4), 250–264.

Becker, G. S. (1964). *Human capital: A theoretical and empirical analysis with special reference to education*. National Bureau of Economic Research.

Becker, G. S. (2007). Health as human capital: Synthesis and extensions. *Oxford Economic Papers, 59*, 379–410.

Becker, G. S., & Tomes, N. (1986). Human capital and the rise and fall of families. *Journal of Labor Economics, 4*(3), 1–39.

Becker, G. S., et al. (2018). A theory of intergenerational mobility. *Journal of Political Economy, 126*(S1), S7–S25.

Berger, T., & Engzell, P. (2019). American geography of opportunity reveals European origins. *Proceedings of the National Academy of Sciences, 116*(13), 6045–6050.

Bjørnskov, C. (2007). Determinants of general trust: A cross-country comparison. *Public Choice, 130*(1–2), 1–21.

Blakemore, S.-J., & Frith, U. (2005). *The learning brain: Lessons for education*. Blackwell.

Borgerhoff Mulder, M. et al. (2009). Intergenerational wealth transmission and the dynamics of inequality in small-scale societies. *Science* (October 30) 326(5953), 682–688.

Borghans, L., et al. (2011). Identification problems in personality psychology. *Personality and Individual Differences, 51*(3), 315–320.

Bourdieu, P. (1980). Le capital social. *Actes de la recherche en sciences sociales, 31*, 2–3.

Brandén, G. (2019). Does inequality reduce mobility? The Great Gatsby curve and its mechanisms. *IFAU Working Paper* 2019:20, Uppsala.

Braun, S. T., & Stuhler, J. (2018). The transmission of inequality across multiple generations: Testing recent theories with evidence from Germany. *The Economic Journal, 128*(609), 576–611.

Burström, K., et al. (2005). Increasing socio-economic inequalities in life expectancy and QALYs in Sweden 1980–1997. *Health Economics, 14*(8), 831–850.

Burt, R. S. (1982). *Toward a structural theory of action: Network models of social structure, perception, and action*. Academic.

Case, A., & Deaton, A. S. (2005). Broken down by work and sex: How our health declines. Ch. 6 in D. A. Wise (Ed.), *Analyses in the economics of aging*, 185–212. Chicago: University of Chicago Press.

Case, A., et al. (2002). Economic status and health in childhood: The origins of the gradient. *American Economic Review, 92*(5), 1308–1334.

Castro-Caldas, A., et al. (1998). The illiterate brain. Learning to read and write during childhood influences the functional organization of the adult brain. *Brain, 121*, 1053–1063.

Clark, G., & Cummins, N. (2015). Intergenerational wealth mobility in England, 1858–2012: Surnames and social mobility. *The Economic Journal, 125*(582), 61–85.

Coleman, J. S. (1990). *Foundations of social theory*. Belknap/Harvard University Press.

Collado, M. D., et al. (2020). *Estimating intergenerational and assortative processes in extended family data*. Working paper, submitted for publication. Accessible at https://janstuhler.files.wordpress.com/2020/10/collado-ortuno-and-stuhler-manucript_v37.pdf.

Commons, J. R. (1924). *Legal foundations of capitalism*. The Macmillan Company.

Conti, G., & Heckman, J. J. (2010). Understanding the early origins of the education–health gradient: A framework that can also be applied to analyze gene–environment interactions. *Perspectives on Psychological Sciences, 5*(5), 585–605.

Conti, G., et al. (2010). The education-health gradient. *American Economic Review, 100*(2), 234–238.

Cook, D. L. (1959). A replication of Lord's study on skewness and kurtosis of observed testscore distributions. *Educational and Psychological Measurement, 19*, 81–87.

Creedy, J. (1985). *Dynamics of income distribution*. Basil Blackwell.

Crow, E. L., & Shimizu, K. (Eds.). (1988). *The lognormal distribution*. Marcel Dekker.

Crystal, S., et al. (2017). Cumulative advantage, cumulative disadvantage, and evolving patters of late-life inequality. *The Gerontologist, 57*(5), 910–920.

Cunha, F., & Heckman, J. J. (2009). The economics and psychology of inequality and human development. *Journal of the European Economic Association, 7*(2–3), 320–364.

Cunha, F., et al. (2006). Interpreting the evidence on life cycle skill formation. Chapter 12 in *Handbook of the economics of education*, vol. 1, 697–812. : North-Holland.

Davison, E., et al. (2015). Brain network adaptability across task states. *PLoS Computational Biology, 11*(1), e1004029.

Deaton, A., & Paxson, C. (1994). Intergenerational choice and inequality. *Journal of Political Economy, 102*(3), 437–467.

Durlauf, S. N. (2002). On the empirics of social capital. *The Economic Journal, 112*, F459–F479.

Durlauf, S. N., & Seshadri, A. (2017). Understanding the Great Gatsby curve. Ch. 4 in *NBER Macroeconomics annual 2017*, 333–393. Cambridge, MA: National Bureau of Economic Research.

Eika, L., et al. (2019). Educational assortative mating and household income inequality. *Journal of Political Economy, 127*(6), 2795–2835.

Fernández, R., & Rogerson, R. (2001). Sorting and long-run inequality. *The Quarterly Journal of Economics, 116*(4), 1305–1341.

Fetter, F. A. (1937). Reformulation of the concepts of capital and income in economics and accounting. *The Accounting Review, 12*(1), 3–12.

Fischer, C. F., et al. (1996). *Inequality by design. Cracking the bell curve myth*. Princeton University Press.

Fisher, I. (1904). Precedents for defining capital. *The Quarterly Journal of Economics, 18*(3), 386–408.

Fredriksson, P., et al. (2016). Parental responses to public investments in children: Evidence from a maximum class size rule. *Journal of Human Resources, 51*(4), 832–868.

Gabaix, X. (2016). Power laws in economics: An introduction. *Journal of Economic Perspectives, 30*(1), 185–206.

Galama, T. J., & van Kippersluis, H. (2013). Health inequalities through the lens of health capital theory: Issues, solutions, and future directions. *Research on Economic Inequality, 21*, 263–284.

Gallup, J. L., & Sachs, J. (2001). The economic burden of malaria. *American Journal of Tropical Medicine and Hygiene, 64*(1), 85–96.

Gibrat, R. (1930). Une loi des répartitions économiques: l'effet proportionnel. *Bulletin de la Statistique Générale de la France, 19*, 469–516.

Glaeser, E. L., et al. (2002). An economic approach to social capital. *The Economic Journal, 112*(483), F437–F458.

Greenwood, J., et al. (2014). Marry your like: Assortative mating and income inequality. *American Economic Review, 104*(5), 348–353.

Grossman, M. (1972). On the concept of health capital and the demand for health. *The Journal of Political Economy, 80*, 223–255.

Grossman, M. (2000). The human capital model. Ch. 7 in A. J. Culyer, & J. P. Newhouse (Eds.), *Handbook of health economics*, Vol. 1. Amsterdam: Elsevier.

Gustafsson, J.-E., et al. (2010). *School, learning and mental health. A systematic review*. The Royal Swedish Academy of Sciences.

Hamilton, K., & Hepburn, C. (Eds.). (2017). *National wealth. What is missing, and why it matters.* Oxford University Press.

Hamilton, K., et al. (2017). Social capital, trust, and well-being in the evaluation of wealth. Chapter 12 in K. Hamilton, & C. Hepburn (Eds.), *National wealth. What is missing, and why it matters*. Oxford: Oxford University Press.

Häuberer, J. (2011). *Social capital theory. Towards a methodological foundation*. VS Verlag für Sozialwissenschaften | Springer Fachmedien.

Heckman, J. J. (2008). Schools, skills and synapses. *Economic Inquiry, 46*(3), 289–324.

Heckman, J. J. (2012). The developmental origins of health. *Health Economics, 21*(1), 24–29.

Heckman, J. J., et al. (2018). Returns to education: The causal effects of education on earnings, health, and smoking. *Journal of Political Economy, 126*(S1), S197–S246.

Helliwell, J. F., et al. (2016). How durable are social norms: . . .? *Social Indicators Research, 128*, 201–219.

Helliwell, J. F., et al. (2018). New evidence on trust and well-being. In E. Uslaner (Ed.), *The Oxford handbook of social and political trust*. Oxford University Press.

Herrnstein, R. J., & Murray, C. A. (1994). *The Bell curve: Intelligence and class structure in American life*. Free Press.

Ho, A. D., & Yu, C. C. (2015). Descriptive statistics for modern test score distributions: Skewness, kurtosis, discreteness, and ceiling effects. *Educational and Psychological Measurement, 75*(3), 365–388.

Horwitz, M. J. (1977). *The transformation of American Law, 1780–1860*. Harvard University Press.

Jordahl, H. (2009). Economic inequality. Ch. 19 in G. T. Svendsen, & G. L. H. Svendsen (Eds.), *Handbook of social capital*. Cheltenham: Edward Elgar.

Kautz, T., et al. (2014). Fostering and measuring skills: Improving cognitive and non-cognitive skills to promote lifetime success. *NBER working paper* 20749. Cambridge, MA: National Bureau of Economic Research.

Kostrubiec, V., et al. (2012). Beyond the blank slate: Routes to learning new coordination patterns depend on the intrinsic dynamics of the learner—experimental evidence and theoretical model. *Frontiers in Human Neuroscience, 6*, 1–14 (Art. 222). https://doi.org/10.3389/fnhum.2012.00222.

Krueger, A. B. (2012). *The rise and consequences of inequality in the United States*. Speech given at Center for American Progress, Washington, DC, January 12.

Lee, S. Y., & Seshadri, A. (2019). On the intergenerational transmission of economic status. *Journal of Political Economy, 127*(2), 855–921.

Limpert, E., & Stahel, W. A. (2011). Problems with using the normal distribution—and ways to improve quality and efficiency of data analysis. *Public Library of Science One, 6*(7), e21403.

Lindahl, M., et al. (2015). Long-term intergenerational persistence of human capital: An empirical analysis of four generations. *Journal of Human Resources, 50*(1), 1–33.

Liu, G. (2011). Measuring the stock of human capital for comparative analysis: An application of the lifetime income approach to selected countries. *Statistics directorate working paper* No. 41. Paris: OECD.

Looi, C. Y. et al. (2016). The neuroscience of mathematical cognition and learning. *OECD education working paper* No. 136. Paris: OECD.

Lord, F. M. (1955). A survey of observed test-score distributions with respect to skewness and kurtosis. *Educational and Psychological Measurement, 15*, 383–389.

Loury, G. (1977). A dynamic theory of racial income differences. Ch. 8 in P. A. Wallace, & A. Le Mund (Eds.), *Women, minorities and employment discrimination*. Lexington, MA: Lexington Books.

Lyon, F., et al. (2015). Introduction: The variety of methods for the multi-faceted phenomenon of trust. Ch. 1 in F. Lyon et al. (Eds.), *Handbook of research methods on trust*, 2nd ed. : Edward Elgar.

McMahon, W. W. (2009). *Higher learning, greater good. The private and social benefits of higher education*. The Johns Hopkins University Press.

McMahon, W. W. (2018). The total return to higher education: Is there underinvestment for economic growth and development? *The Quarterly Review of Economics and Finance, 70*, 90–111.

Micceri, T. (1989). The unicorn, the normal curve, and other improbable creatures. *Psychological Bulletin, 105*, 156–166.

Mincer, J. (1970). The distribution of labor incomes: A survey with special reference to the human capital approach. *Journal of Economic Literature, 8*(1), 1–26.

Mitzenmacher, M. (2004). A brief history of generative models for power law and lognormal distributions. *Internet Mathematics, 1*(2), 226–251.

Newman, M. E. J. (2005). Power laws, Pareto distributions and Zipf's law. *Contemporary Physics, 46*(5), 323–351.

O'Rand, A. M. (1996). The precious and the precocious: Understanding cumulative disadvantage and cumulative advantage over the life course. *The Gerontologist, 36*(2), 230–238.

Ostrom, E., & Ahn, T. K. (2009). The meaning of social capital and its link to collective action. In G. T. Svendsen & G. L. H. Svendsen (Eds.), *Handbook of social capital*. Edward Elgar.

Pareto, V. (1897). *Cours d'Économie Politique*. Rouge.

Poortinga, W. (2002). Social capital: An individual or collective resource for health? *Social Science & Medicine, 62*(2), 292–302.

Putnam, R. (2000). *Bowling alone*. Simon & Schuster.

Putnam, R. (2001). Social capital: Measurement and consequences. *Canadian Journal of Policy Research, 2*, 41–51.

Rice, T. W., & Feldman, J. L. (1997). Civic culture and democracy from Europe to America. *The Journal of Politics, 59*(4), 1143–1172.

Rothstein, B. (2011). *The quality of government. Corruption, social trust and inequality in an international perspective*. University of Chicago Press.

Scheffer, M., et al. (2017). Inequality in nature and society. *Proceedings of the National Academy of Sciences, 114*(50), 13154–13157.

Schultz, T. W. (1962). *Investment in human beings*. University of Chicago Press.

Scrivens, K., & Smith, C. (2013). Four interpretations of social capital: An agenda for measurement. *Statistics directorate working paper* No. 55. Paris: OECD.

Simon, H. A. (1955). On a class of skew distribution functions. *Biometrika, 42*, 425–440.

Smith, A. (1776). *An inquiry into the nature and causes of the wealth of nations*. Several later editions.

Solow, R. M. (1957). Technical change and the aggregate production function. *Review of Economics and Statistics, 39*(3), 312–320.

Sporns, O. (2011). *Networks of the brain*. MIT Press.

Uslaner, E. (2008). Where you stand depends upon where your grandparents sat. *Public Opinion Quarterly, 72*(4), 725–740.

Uslaner, E. (Ed.). (2018). *The Oxford handbook of social and political trust*. Oxford University Press.

Uslaner, E., & Brown, M. (2005). Inequality, trust, and civic engagement. *American Politics Research, 33*(6), 868–894.

Vosters, K., & Nybom, M. (2017). Intergenerational persistence in latent socioeconomic status: Evidence from Sweden and the United States. *Journal of Labor Economics, 35*(3), 869–901.

Warrington, W., et al. (2000). Student attitudes, image and the gender gap. *British Educational Research Journal, 26*(3), 393–407.

Weil, D.N. (2014). Health and economic growth. Ch. 3 in P. Aghion, & S. Durlauf (Eds.), *Handbook of economic growth*, Vol. 2B, pp. 623–682.

Yule, G. U. (1925). A mathematical theory of evolution based on the conclusions of Dr. J. C. Willis. *Philosophical Transactions of the Royal Society of London B, 213*, 21–87.

Chapter 5
Interaction as the Source of Inequality

In the preceding two chapters, the emergence and development of inequality were studied from the perspective of an individual. Obviously, any individual lives and grows in a societal context, but the environment with which an individual interacts has so far been described as relatively anonymous and abstract. The focus will now shift to the process of interaction itself, which in many situations is a source of inequality.

Interaction may take many different forms—imitation, cooperation, competition, among others. The situations underlying these behavioural strategies are universal and can be found in many species. The strategies observed among humans are consequently to some extent the result of biological evolution in response to recurrent, basic problems of survival. The structure and character of solutions depend heavily on the parameters of the situation, such as group size and degree of symmetry, and so will the resulting level of inequality.

5.1 Bilateral Relations

The smallest imaginable group consists of two parties. This may be two individuals such as Robinson Crusoe and Friday on an otherwise desert island, but may equally well refer to two groups that for the sake of analysis can be considered as decision-makers—households, companies, political parties, or states. The process by which a group reaches a common position may be complex but falls outside the analysis.

Looking across human history in a broad geographic perspective, inequality has varied substantially.[1] As explained earlier,[2] two of the main drivers of inequality are the characteristics of the main categories of wealth—material (physical and

---

[1] A general source-book on human prehistory is Scarre (2018).

[2] Section 3.1.

financial), human and social capital—and the degree to which this wealth can be transmitted from one generation to the next. If physical capital such as cattle or agricultural land is relatively important and can be transmitted successfully across generations, the probability of stark inequalities is high.

Agricultural Societies

In agricultural societies, material wealth, particularly in the form of land, dominates and is also highly heritable.[3] This explains in part the high and persistent levels of inequality observed in societies that rely on intensive agriculture. The relationship to arable land tends to be the defining basis of social inequality in these societies, as expressed in the categories of landlords, smallholders, tenants and wage labour, to which can be added serfs and slaves where such have existed.

As illustrated by Fig. 1.1, inequality in agriculturally dominated Europe increased at least from 1300 CE well into the industrialised phase, interrupted only by the Black Death in the mid-fourteenth century and its aftermath. In Scheidel's perspective, it is the exceptional mortality figures in the mid-fourteenth century that explain the decrease in inequality essentially via shortage of labour.[4] But as Alfani points out, there were epidemics in the seventeenth century that were regionally equally mortal but did not generate a decrease in inequality.[5]

The conclusion seems to be that norms and institutions play a part in the explanation of inequality dynamics. Institutional arrangements have added to inequality via inheritance legislation and prohibition of collective action on the part of the less wealthy. Adam Smith summarises the imbalanced situation between the two main categories in late eighteenth-century Britain in a passage in *The Wealth of Nations*:

> What are the common wages of labour, depends everywhere upon the contract usually made between those two parties, whose interests are by no means the same. The workmen desire to get as much, the masters to give as little, as possible. The former are disposed to combine in order to raise, the latter in order to lower, the wages of labour.
>
> It is not, however, difficult to foresee which of the two parties must, upon all ordinary occasions, have the advantage in the dispute, and force the other into a compliance with their terms. The masters, being fewer in number, can combine much more easily: and the law, besides, authorises, or at least does not prohibit, their combinations, while it prohibits those of the workmen. We have no acts of parliament against combining to lower the price of work, but many against combining to raise it. In all such disputes, the masters can hold out much longer. A landlord, a farmer, a master manufacturer, or merchant, though they did not employ a single workman, could generally live a year or two upon the stocks, which they have already acquired. Many workmen could not subsist a week, few could subsist a month, and scarce any a year, without employment. In the long run, the workman may be as necessary to his master as his master is to him; but the necessity is not so immediate.[6]

---

[3] Shenk et al. (2010).

[4] Scheidel (2017).

[5] Alfani (2021).

[6] Smith (1776), Book I, Chapter VIII.

As observed by Smith, the source of asymmetry is not only the difference in wealth but also in number, which makes organisation much more difficult among workers, quite apart from the legal prohibition against such initiatives. Together with the difference in time horizons, these factors create a stark imbalance between the bargaining positions of the two parties.

The fallback position, which materialises if the two parties are unable to reach a modus vivendi, is obviously of paramount importance to the outcome, as Smith observes. In a classical study of oriental despotism, Wittfogel argued that centralised management of the irrigation system necessary for agricultural production facilitated the emergence of authoritarian regimes.[7] Wittfogel's hypothesis has been criticised but nonetheless offers a useful perspective on the power balance between the elite and the majority in a society where there are few or no exit alternatives.[8] Emigration, when possible, has been an efficient threat to strengthen one's bargaining position. During the second half of the nineteenth century, emigration to the United States was a viable option in European countries that were hit hard by food supply crises, such as Ireland and Sweden.[9] Among those who remained in their country of origin, political organisation was facilitated by the demographic crisis.[10]

Power imbalances may persist across generations as a stable social equilibrium, in which a large part of the landholdings is controlled by a small elite and the majority of the population live near the existence minimum, either as small landowners or as propertyless agricultural labour. This unequal distribution of wealth may be the source of economic inefficiencies, in the sense that investment projects that would raise agricultural productivity are not carried out. It is well known that productivity in small farms is generally higher than in large farms in agricultural economies,[11] but credit constraints and the necessarily higher risk aversion among less wealthy producers prevent productivity-enhancing investments from being implemented.[12] Even though tenants and labourers could in principle be able to buy land and finance their purchase with increased production, the land may not be for sale because current owners see traditional holdings as a sign of status and power, or because alternative investment may be considered too risky.

This diagnosis is supported by an analysis carried out by the World Bank on the connections between inequality of land holdings and macroeconomic growth rates. The Bank studied a group of countries that had made the transition from agricultural to industrialising economies during the period 1960–2000 and found a strong positive correlation between the equality of land holdings (the principal asset) in 1960 and the average growth rate during the four decades that followed.[13] The above

---

[7] Wittfogel (1957).

[8] See the survey by Mitchell (1973).

[9] Akenson (2011).

[10] Karadja and Prawitz (2019) provide detailed evidence for Sweden.

[11] See, for example, Lipton (2009), Chap. 2.

[12] Bardhan et al. (2000).

[13] World Bank (2005), p. 163.

mechanisms, together with negative effects of inequality on human capital investments, social cohesion and the workings of the political system, bolster the idea that
the relationship between land distribution and GDP growth is causal.[14] Further, the
overall distributional effect of growth on inequality is correlated to the initial level of
inequality: the higher the level of inequality at the start of a period, the less efficient
is the economy in spreading the effects of economic growth to the lower income
strata.[15]

In summary, a high degree of inequality in an agricultural society may generate a
development trap, characterised by slow growth and persistent inequality.

Industrialised Societies
During the transition to an industrialised society, the main dividing line in society
gradually changed from landlords versus labour to industrial capital versus labour,
but the central issue remained what David Ricardo formulated as the core of political
economy—how to divide the value added in the production system between labour
and capital.[16] Over time, the borderline between capital-owners and labour has
become less clear. While part of the population in industrialised countries still
have limited or no capital assets, the middle class have assets in the form of housing
and shares in pension funds, and large capital owners often have large wage earnings
alongside their capital earnings.[17] The effects of the increased spread of sharehold-
ing should not be exaggerated, however. Empirical studies of corporate ownership
show that, across the world, both ownership and control are still concentrated.[18] The
categories of capital and labour are still relevant.

The tax system is important. Reduced progressivity in income tax rates will lead
to higher salaries among high-income earners, implying an increased wage share at
the same time as inequality increases.[19]

In a general sense, the relationship between capital owners and labour has become
more abstract. The classical image of a group of employees meeting the local mill
owner is outdated. Negotiations are often held at a higher level than the workplace
and decided for a collective rather than for an individual. An employee who is
dissatisfied with employment conditions can discuss and negotiate, but can also
choose to leave for another employment without any preceding negotiation, in which
case the bargaining process is tacit.

As a consequence, there is no obvious apriori link between inequality and wage
share. Empirically, it exists, however; there is a strong, positive connection between
capital shares and inequality, which has grown stronger over the past century. The
link is stronger at the top of the distribution and is also stronger when top capital

---

[14]Deininger and Olinto (2000).

[15]World Bank (2005), p. 87.

[16]Ricardo (1821).

[17]Krueger (1999).

[18]La Porta et al. (1999) (global), Claessens et al. (2002) (Asia), Faccio and Lang (2002) (Western
Europe), Ireland (2005) (UK and U.S.).

[19]Rubolino and Waldenström (2020).

incomes predominate.[20] There are consequently good reasons why the issue of wage shares has continued to attract interest among researchers.

Measuring capital is difficult. Capital can be more or less easy to liquidate and thereby confers more or less economic power. Claims on future pensions cannot be used as collateral for loans and therefore represent a fairly abstract form of asset. Most analyses therefore limit their definition of capital to marketable forms.[21]

Globally, the picture is somewhat diverse, but there has been a general tendency towards falling wage shares during the last four to five decades in advanced, emerging-market, as well as developing economies.[22] In some countries and sectors, the decline of the wages share has been particularly strong: in US manufacturing, it fell from 0.61 to 0.41 between 1967 and 2007.[23] Part of the explanation lies in factors under the heading of *globalisation*. The expansion of a global system of communication has integrated financial markets in different parts of the world and also made it possible to decentralise production while maintaining control at the centre.[24] In other words, the spectrum of possibilities open to capital has expanded more rapidly than that of labour, rendering the fallback position of capital relatively more favourable in the (open or tacit) bargaining with labour. With rising levels of education in emerging and developing economies, this increased competition also affects skilled labour. The development of information technology has made it possible to purchase qualified services across large distances.

Other factors have contributed. Lowered prices on investment goods have contributed to *increased capital intensity* and labour productivity, and wage development has not followed productivity.[25] The *concentration of market shares* in the United States has increased, which pushes the labour market in the direction of increasing monopsony.[26]

A debate in recent years of direct relevance to the problem of wage and capital shares concerns the relation between capital payoff (r) and the general growth rate (g) of the economy, that is, whether r is greater than or less than g.[27] An investigation of 16 advanced economies during the period 1870–2015 lends some support to the hypothesis that r is greater than g.[28] The average payoff to risky investments (estate property, shares) has been high and relatively stable in the interval 6–8 per cent, whereas payoff to so-called "risk-free" investments such as government bonds has

---

[20] Bengtsson and Waldenström (2018). More generally, there is a strong correlation between top income shares and inequality as measured by the Gini coefficient; see Sect. 5.6. On the role of top incomes, see also Atkinson et al. (2011).

[21] Davies and Shorrocks (2000).

[22] IMF (2017), Chap. 3.

[23] Kehrig and Vincent (2017).

[24] Baldwin (2016, 2019).

[25] Karabarbounis and Neiman (2014), Bassanini and Manfredi (2014), Kehrig and Vincent (2017).

[26] Autor et al. (2017).

[27] Piketty et al. (2013), Krusell and Smith (2015).

[28] Jordà et al. (2019).

varied between 1 and 3 per cent and has been surprisingly volatile. Strong capital accumulation in recent decades in combination with stable payoffs explains in part the rising capital shares in the economy.

This does not elevate the inequality between r and g to the status of a general law. During large parts of the preindustrial era, g was in the neighbourhood of zero, but the evidence to support the idea that a positive r explains rising inequality is weak.[29]

Countervailing Forces: Education and Organisation

In a secular time-perspective, the period between the late nineteenth century and the 1970s represents a deviation from the general trend of increasing inequalities. Two major wars led to massive destruction of physical capital in countries that were directly hit by military operations, and the Great Depression had a similar effect on financial capital in a broader category of countries. Both had a measurable and equalising impact on the distribution of wealth. But more generally, the origin of changing labour shares during this period must at least partly be sought outside the economic sphere in the narrow sense.

Education is an obvious factor of importance to equality, given the strong link between skills and wages. The development of mass education systems in Europe during the nineteenth century led to increased literacy and numeracy in populations, starting in western Europe with Japan and eastern Europe following suit and the rest of the world taking off only after the Second World War. Because of the necessarily gradual expansion of educational systems, inequality first rises and then falls in the transition from illiteracy to literacy. Human capital inequality peaks when countries reach a level of 3–4 average years of schooling.[30] At the global level, human capital inequality peaked during the second half of the twentieth century.

The picture is somewhat different between the early developers and the world at large, however. While the expansion of general education has contributed to equality at the global level, it has been difficult to find similar equalising forces in the expansion of higher education in the most advanced countries. The inequality in transitions to higher education among social strata illustrated for instance in Fig. 3.2 has proven resistant to change.[31]

At present, there is a discrepancy between the position of western Europe and the United States in the world income distribution and the corresponding position in the distribution of human capital, indicating that a shift to the benefit of Asian countries can be expected in the coming decades.[32]

Turning to the specifically social and political arena, Adam Smith observed in the passage from *The Wealth of Nations* quoted above that both the *capacity to organise* and the *legislative framework* can be expected to affect the balance between labour and capital. Two factors of major importance changing the landscape from the late

---

[29] Alfani (2021).

[30] Morrisson and Murtin (2013).

[31] Bar Haim and Shavit (2013) (22 European countries plus Israel and Turkey).

[32] Morrisson and Murtin (2013).

nineteenth century onwards were *trade unions* and *general suffrage*. There has been a positive correlation between trade union density and the wage share in the past 50 years in advanced capitalist countries.[33] The effect is far from linear, however, and it differs across countries due to differences in both history and levels of unionisation. In countries with a high union density, unions tend to moderate their wage demands. At the same time, unions contribute to a general reduction in earnings inequality. There are indications that decreasing union densities in recent decades have contributed to the falling wage share.[34]

As for the purely political aspect, general suffrage can contribute to increased wage shares and lower inequality via a number of channels—expanding *systems of education* and increased *political participation* in all countries, *minimum wages* in some countries, and so on. Controlling for a number of possibly relevant determinants, there is a robust and statistically significant correlation between the degree of democracy in a country and the level of manufacturing wages. This holds both across countries and over time within one and the same country.[35]

Positional Goods

In standard economic models of consumption, agents are assumed to base their evaluation of their current state on their own consumption only. What other individuals consume is irrelevant, both in the positive (altruism) and the negative (jealousy and spite) sense. This is obviously not correct. Comparisons are central to the functioning of a market economy, for instance when it comes to re-allocation of capital to more profitable activities, or to changes of employment. So it is logical that comparisons also play a role when evaluating one's own consumption level.

While the general standard of living in the world has risen as a result of increased production levels, there is a category of goods and services that is not supplied in increasing quantities, what is referred to as *positional goods*. Property in attractive living areas and leading positions in business organisations or political institutions are examples of this category. The relative income or consumption level is a more abstract example of the same category: only one per cent of the population can belong to the uppermost percentile group.

In *Social Limits to Growth*, Fred Hirsch made the point that as the general level of consumption of ordinary goods grows in a society, the relative importance of positional goods will increase.[36] Empirical research has confirmed Hirsch's view; relative levels of consumption are important.[37] There are significant differences between various types of goods and services; some are more positional than others.

The importance of positional goods to the analysis of inequality is that it gives rise to a component of inequality that is, in a sense, ineradicable. There is also a risk

---

[33] Bengtsson (2014).

[34] For an analysis of the latter two statements, see Visser and Checchi (2009).

[35] Rodrik (1999). See further Sect. 5.6.

[36] Hirsch (1976).

[37] For an example, see Carlsson et al. (2007).

that wealthier strata, once their basic needs are satisfied, will focus an increasing share of their attention and capital on positional assets, which may have negative effects on the workings of the economy.[38]

5.2   Discrimination and Aggression

As pointed out in Chap. 3, individuals are not in general remunerated purely on the basis of their human capital in the labour market. Arbitrary factors such as ethnic origin and gender distort the relationship between human capital and wages. In the present section, discrimination and aggression will be analysed at the collective level, as part of a general struggle for dominance over resources and positions.

The binary opposition between different social groups analysed in the previous section emanates from positions defined within the economic sphere, on the basis of property or dominant income source. Exchange patterns between individuals and between groups are also affected by external categories that need not a priori interfere with economic relationships, such as gender, ethnic origin or class. In practice they do, however; such categories are strong predictors of economic outcomes. They intersect with economic categories, and correlations are often strong, which makes them an important source of inequality.

Most forms of discrimination go far back in time, and some are archaic. The defining characteristic is always easily visible or audible—sex, skin colour, handicaps, language, and so on. No discrimination, as far as is known, is based on blood groups.

Discrimination, in a basic sense, is irrational. Making an economic decision dependent on an arbitrary sign goes against the rationality norm. Some have tried to solve this problem in theory building by adding a preference dimension, which makes the theory tautological.[39] Others ascribe discriminatory decision-making to uncertainty, for instance about productivity when hiring an employee, and classify this form of behaviour as statistical discrimination. But statistical discrimination is also discrimination.

Gender
Discrimination by *gender* is old and universal, often sustained and deepened by social and religious norm systems. The status of women depends on their economic importance to the social household. In some cases, the development of women's status can be followed chronologically in archaeological strata. Archaeological dietary data from China (based on isotope analysis of human and animal bone

---

[38] We will return to this problematique in Sect. 6.5.

[39] Arrow (1998). Friedman represents this quasi-tautological view: "The man who exercises discrimination pays a price for doing so. He is, as it were, 'buying' what he regards as a 'product'. It is hard to see that discrimination can have any meaning other than a 'taste' of others that one does not share." (Friedman 1962, Chapter VII).

samples) indicate a high degree of equality during the Neolithic period, when the contribution of men and women to the household economy were comparable, and also in early farming societies (5000–2900 BCE).[40] The introduction of wheat and barley as well as domesticated herbivores during the Late Neolithic period (around 2600–1900 BCE) and during the Bronze Age Eastern Zhou Dynasty (771–221 BCE) led to a change that affected females more than males. Consumption of animal products was reduced and became increasingly unequal between the sexes. A higher rate of skeletal stress markers among females together with increased body height differences indicate increasing inequality between the sexes, which is supported also by the increased wealth of male burials.

The current situation at the global level reflects the notion that the position of women is determined by a combination of material circumstances and cultural factors. Gaps between men and women exist in education, health, and economic and political opportunity.[41] Average female labour force participation (FLFP) hovers around 50 per cent, but with levels and trends varying across regions. With respect to GDP per capita, the FLFP displays a U-shaped dependence: values are high in low-income Sub-Saharan Africa and in OECD countries, while they are particularly low in the Middle East and North Africa. In low-income countries, a high value of FLFP reflects the need to work to support the household, in the absence of social protection programmes. As income rises and social protection improves, women tend to withdraw from the market, in favour of unpaid work in the household. At higher income levels, the FLFP increases again, primarily as a result of better education, lower fertility rates and political emancipation.

The gap between the sexes in labour force participation has been shrinking since the 1990s, but relatively slowly. Variations are also large within the OECD group, with small gaps in the Nordic countries and a large one in Japan.

The gender-based wage gap for the same occupations is still significant when controlling for individual characteristics, such as education. Occupational segregation and fewer working hours, in combination with differences in work experience, explain on average about 30 per cent of the wage gap. As pointed out in Sect. 3.4, the wage gap increases steeply during childbearing and childrearing years. The child penalty is estimated at 14 per cent across the OECD countries. Wage gaps in middle-income countries vary significantly. They are lower in countries where only a limited share of the female population is employed, and where women are then often more highly educated than their male colleagues.

Ethnic Discrimination

Discrimination on an ethnic or cultural basis—skin colour, language, caste, etcetera—can be analysed along similar lines.[42] Wage gaps can be explained to a varying extent

---

[40] Dong et al. (2017).

[41] The IMF Staff surveys, Elborgh-Woytek et al. (2013) and Stotsky et al. (2016), are the sources for the data that follow.

[42] For a joint survey of research on race and gender issues in the labour market, see Altonji and Blank (1999) (U.S. data).

by differences in skills, experience and position, and the remainder—the unexplained variance—is often taken to be a sign of discrimination. A few caveats are necessary here. If appointments in the labour market are affected by discrimination, then statistical analyses that control for positions and seniority will underestimate the actual level of discrimination. Further, if personal choices by individuals belonging to racial minorities, the quality of schooling, etcetera are affected by prevailing discrimination patterns, that will also contribute to an underestimate of the actual level of discrimination. Social rank affects health and performance,[43] so discrimination can lead to self-fulfilling statements that minorities are low performers. On the other hand, variables used in statistical analyses are often crude approximations of actual skills, which creates a risk of overestimating discrimination.

With these caveats in mind, a number of studies confirm the hypothesis that discrimination is important, but that the character and the degree vary with location and history. The significance of the geographical factor is illuminating, as it signals the persistence of collective factors that go beyond what is objectively relevant to decision-making. It has been shown that ethnic minorities with a higher education in the US suffer from a wage gap vis-à-vis the white majority.[44] For African American men, only about a quarter of the wage gap is explained by factors such as formal education and English language proficiency. For a subsample born outside the South to parents with some college education, these factors account for the entire wage gap. This indicates the presence of multiple social equilibria, where history is decisive for the outcome.

The Indian caste system offers another concrete illustration of the negative effect of discrimination both on human capital formation and productivity, given human capital assets. Making caste identity public in North Indian classrooms has proven to reduce the cognitive performance of low-caste boys by 23 per cent. Discrimination between higher-caste landowners and lower-caste tenants in access to groundwater for irrigation has reduced the latter's agricultural yields by 45 per cent.[45]

The Role of Aggression

Not only visible traits such as ethnic origin or gender can play a role in a social struggle for material benefits and positions. Behaviour is central to the outcome of social interaction and has been studied by ethologists in many different species. Behaviour can also be fixed in personality traits, because this may improve reproductive success.[46] *Aggression* is of particular interest, as it may generate benefits to the bearer of this trait, even if this individual does not have a productivity or fitness above average. Aggressive personalities have been studied extensively among ethologists, and have been found to be of relevance to rank.[47] Aggressiveness

---

[43] Abbott et al. (2003), Sapolsky (2004).

[44] Black et al. (2006).

[45] Hoff (2016).

[46] Wolf et al. (2007).

[47] See the survey by Holekamp and Strauss (2016), and, concerning mating, Cowlishaw and Dunbar (1991). On specific species, see for instance Camerlink et al. (2015) (pigs), McEvoy et al. (2013) (lizards), and Favati et al. (2017) (domesticated fowl).

predicts future rank well compared to other personality traits and also better than experience from previous contests.

There is a priori no reason to doubt that similar factors also play a role in human populations. It is important, however, to separate between different forms of aggression. One variant indicates a lack of control, is generally dysfunctional, and tends to hurt both the subject and the group.[48] Strategic aggression, by contrast, is a signal that need not result in the practice of violence but serves the purpose of strengthening the position of the subject when confronting other individuals in conflicts over resources or positions. It is the second form that is of interest as a source of inequality.

5.3 Conformity and Imitation

The standard model of rational decision-making lacks realism. Human beings constantly make decisions of varying importance, and a full analysis of all available alternatives with respect to their consequences in all these situations is simply not feasible. Further, consequence analyses, when they are made, always suffer from some uncertainty, which must be handled. The normal solution in everyday life to this insurmountable decision problem is to resort to thumb rules. Common thumb rules are to stick to the solution one made previously in a similar situation—which gives to the status quo a natural primacy—or to follow the example of others.

Imitation
These sorts of thumb rules are often surprisingly efficient, in spite of their brutal simplicity.[49] The inertia of the status quo lies behind the establishment of habits and norms, some of which contribute to maintaining efficient behavioural equilibria in daily life—which is not to say that all equilibria are efficient. Imitation is also present in preference formation, for instance when it comes to fertility choices.[50]

Simplicity comes at a price, though, because in some situations the drive to conformity is dysfunctional. In classical experiments by Sherif and by Asch, subjects yielded to group pressure instead of relying on their own perceptions.[51] What is needed for an imitative strategy to be successful is consequently a balance between autonomous thinking and sensitivity to group judgements.

There are indications of herd behaviour in financial markets.[52] This shows that complex algorithms do not insure against outcomes that are adverse both to users and to the system.

---

[48] Provençal et al. (2015).

[49] Gigerenzer and Todd (1999).

[50] Hensvik and Nilsson (2010), Buyukkececi et al. (2020).

[51] Sherif (1936), Asch (1956). See also Janis (1982).

[52] Bikhchandani and Sharma (2000), Chang et al. (2000),

Imitation or herd behaviour is linked to inequality in different ways. Education, self-confidence and critical thinking reduce the risk of yielding to inefficient or destructive herd reflexes, so persons with a short education or low status are generally more vulnerable. Education or status is no guarantee against herd behaviour, however, as shown by many counterexamples. The Nazi regime in Germany received strong support from highly educated groups such as lawyers and physicians.[53] Above the individual level, neighbourhoods, cities or whole countries may end up in stable social equilibria that have far-reaching negative effects on well-being. The reason is that a decision rule that is contingent on the behaviour of others is indeterminate; in general, more than one outcome is possible when a small or large group of interacting individuals apply this rule.

School-children and students are affected by others belonging to the same group. Such *peer effects* may be difficult to estimate,[54] but the general view is that they exist.[55] They may be positive[56] or negative[57] when measured by performance, so the effect on inequality is ambiguous.

Segregation

Segregation can have different bases and assume different forms. Normally, spatial segregation is in focus, specifically in its residential form. In principle, segregation need not imply adverse effects for any of the parties involved, but it often does. Standard indicators used for measuring differences in outcomes—human capital formation, incomes, health, and crime rates—testify to the adverse effects of segregation.

In the United States, which is one of the most widely studied countries, the character and form of segregation shifted markedly during the twentieth century.[58] Into the 1960s, segregation was primarily racial and based on spatial separation of whites and blacks. Gradually, segregation at the level of states and counties gave way to segregation at the level of cities and neighbourhoods. During the last three decades of the twentieth century, residential segregation increasingly became based on class rather than on race or ethnicity.

Strong neighbourhood effects have been recorded across areas in the United States, measured by *income mobility* between generations.[59] Upward mobility is fostered by low residential segregation, low income inequality, good primary schools, high levels of social capital and family stability. The spatial dimension turns out to be decisive: both black and white children growing up in areas with large African American populations have lower rates of upward income mobility. Individual parameters are

---

[53] On the legal profession, see Stolleis (1994) and Joerges and Ghaleigh (2003), on the medical profession, see Hanauske-Abel (1996) and Pross (1991).

[54] Angrist (2014).

[55] Sacerdote (2011).

[56] Boucher et al. (2014).

[57] Winson and Zimmermann (2014).

[58] Massey et al. (2009).

[59] Chetty et al. (2014).

still important, but the significance of spatial location indicates that certain community equilibria may have adverse effects on life prospects.

The effect of segregation on *health* is complex and somewhat mixed.[60] Isolation segregation in the United States correlates with poor pregnancy outcomes and high mortality for blacks, but there are indications of certain protective effects of living in clustered neighbourhoods, compensating for social and economic consequences at large. Such effects may explain the general tendency for migrants to gravitate towards neighbourhoods that are populated by groups with a similar ethnic background, irrespective of the negative socioeconomic prospects that are associated with such areas. Such a tendency—*homophily*—can leave traces at the aggregate level in the form of pronounced segregation without being particularly strong in itself.[61]

Trust

At the regional or national level, differences in social equilibria can be identified with respect to *trust*. Even within the relatively homogeneous group of high-income OECD countries, substantial differences in levels of general trust are measured.[62] There is a stable relationship between equality and trust, both at the national and the regional level,[63] and it is tempting to couple the high trust levels in the Nordic countries to the correspondingly high levels of equality. This may be part of the explanation, but there is reason to believe that the roots of social trust in these countries are deeper and go far back in time. When trust levels are measured in the United States, the best single migration-based positive determinant of social capital is the fraction of the population that is of Scandinavian descent.[64] But Scandinavians migrated in large numbers to the United States 150 years ago, when the Scandinavian countries were no forerunners in equality. Nordic welfare states and high equality levels are a relatively recent phenomenon. At least some of the explanation should probably be sought in deep patterns of family formation, where north-western Europe differs from the rest of Europe. In the agricultural economy, young people often worked several years far from home before forming a family, and the social network thus formed was much wider and less family-centred than in central and southern Europe.[65] An important fact is that young people while working as servants kept the social status of their family of origin and were not considered to be socially inferior to the master for whom they worked.

No matter what the origin of current patterns is, a general conclusion from these examples is that populations of various sizes may be trapped in social equilibria that are characterised by lower levels of well-being than are compatible with prevailing material conditions. *Inside* each population, there may be more or less pronounced

---

[60] Kramer and Hogue (2009).

[61] See Sect. 6.3.

[62] A general source book on social and political trust is Uslaner (2018). See also Hamilton et al. (2017) and Sect. 4.5 for further references to the literature.

[63] See Gustavsson and Jordahl (2008) for the regional level, Section 4.5 in general.

[64] Putnam (2001).

[65] Hajnal (1982).

inequality, but an important source of inequality is the difference at the level of populations.

Single individuals may realise the inefficiency of a prevailing equilibrium but are at the same time unable to change the state of affairs, given the collective nature of the outcome.

5.4   Networks

Human life is embedded in networks—natural, social as well as technical. Interdependence in large ecosystems can successfully be described as networks. Human social structures—families, groups, companies, societies—can all be viewed and analysed within the conceptual frame of networks. Large infrastructure networks for energy supply or telecommunications are preconditions for an industrialised society.[66]

In essence, a network is a set of points—nodes, agents, etcetera—united by a number of links. In transport or electricity networks, vehicles and energy flow through the system, and nodes may be crossings or transformer stations. In social networks, links are often less visible, being constituted by acquaintance, affective bonds or norms. Sometimes the two are intertwined, for instance in the management of water supply systems.[67]

Applications to social systems exist in a variety of fields:

- *Learning and diffusion of knowledge*: spread of language and culture, scientific citations, diffusion of innovations[68]
- *Health-related processes and habits*: disease transmission,[69] smoking,[70] sexual contacts[71]
- Informal *insurance systems*[72]
- *Markets* for labour[73] or goods and services[74]
- *Crime* and *mobbing*.[75]

---

[66] A broad, popular survey is Barabási (2002), whereas Newman (2003) is more technical. Jackson (2008) and Easley and Kleinberg (2010) are comprehensive tutorials. Specifically, on ecological networks, see Ings et al. (2009). Boissevain (1974) is an early survey of the social science literature.

[67] Ostrom (1992).

[68] Strang and Soule (1998).

[69] Jackson (2008), Sect. 7.2.

[70] Christakis and Fowler (2008).

[71] Liljeros et al. (2001).

[72] Bloch et al. (2008).

[73] Granovetter (1973).

[74] Kirman (1983), Kirman and Vriend (2000) (Marseille fish market), Uzzi (1996) (New York apparel industry).

[75] Ballester et al. (2006), Boguslaw (2017).

What a network approach adds to more general descriptions of social phenomena is precision and concreteness, at the price of complexity and problems with data supply and quality. In many situations, the increased level of complexity is necessary in order to understand the emergence of qualitatively new behaviour.

Given the diversity of fields of application, it is risky to generalise about the properties of networks, but with this caveat in mind, a few recurrent observations may nonetheless be put forward. The emergence and history of a network is obviously of great importance to its structure and characteristics. There are substantial differences between technical networks, which are designed from above under strict budgetary and efficiency conditions, and social networks, which grow more or less spontaneously. Issues of capacity, congestion and routing are vital to the functioning of technical networks. Maintaining a social contact also requires time and energy, but in this case capacity constraints are felt more at the nodes than in the links.

There are both national and class-related differences in the character of social networks. A common conception is that in modern societies, kin-based ties are weakening while professional and other ties are increasingly important in social networks. This holds true mainly for the Northwest-European cultural area and the New World countries that share their cultural heritage. In southern and eastern Europe, close kin relations are still maintained.[76] This divide, logically, follows the same line as the trust differences reported above.[77]

Within countries, there are tendencies not only that higher strata have higher levels of social capital, but also that networks are to a higher degree associational (formal social capital).[78] In countries with high levels of inequality these differences between classes are further magnified. Informal contacts are less clearly stratified by class.

Inequality in Networks

Inequality is visible in networks in a number of different dimensions. The most obvious indicator is that the number of links attached to a node—the *degree* of the node—varies significantly across the network. The range of variation depends on the growth history of the network, but even under assumptions that favour equality, distributions will be skewed towards higher values. Randomly assigned links from incoming new nodes yield a more even distribution than links assigned on the basis of preferences for nodes that already have many links.[79] In practice, these ideal forms are seldom observed; the typical network is a hybrid.

Power distribution and inequality is not a function of degree only. Who is a *key player* is determined both by degree and by position in the network.[80]

---

[76] Höllinger and Haller (1990).

[77] Hajnal (1982).

[78] Pichler and Wallace (2009).

[79] See Sect. 6.4.

[80] Ibid.

Clustering is another indicator of importance. The underlying question in this case is, given a node that has links to two other nodes, what is the probability that these two other nodes are connected? A high probability indicates a high level of clustering, which in turn has consequences for the structure and functioning of the network. Segregation represents a high degree of clustering visible at the aggregate level. The tendency to form links with agents that are in some respect similar to oneself—*homophily*—is sometimes part of the explanation for segregating patterns.[81]

The concept of *assortativity* connects degree and clustering. If high-degree nodes show a tendency to be connected to other high-degree nodes, the network in question exhibits positive assortativity. There seems to be a tendency for technical networks to be negatively assortative, in which case one talks of a *hub-and-spokes* structure, which is common in large-scale logistic systems. In social systems, positive assortativity indicates a hierarchic structure, with obvious consequences for the distribution of power within the network.

Learning in Networks

Learning processes in networks are also important to power distribution and inequality. Where learning takes place on the basis of contacts with other agents in a network, the level of knowledge and the accuracy of information will be highly dependent on the structure of the network. In a typical population, individuals will base their judgements on a combination of prior knowledge and incoming information.[82] Depending on this mix and the way information is processed, a population may end up in a low-accuracy equilibrium. This holds even if decision-making is rational (Bayesian).[83] It is even possible for a rational, autonomous observer who receives correct information—untainted by faulty signals in the environment—to end up with a consistent but false image of reality.[84] The risk that this scenario will develop is obviously greater for individuals with a poor information base. A powerful agent in a network naturally has a greater possibility of pushing the learning process in the desired direction than a weaker one.

A different situation prevails when the agents in a network are to form a common judgement, for instance whether a certain statement is correct or false. They may exchange information and arguments across the network but are to form their own opinions once this exchange is over. A classic result from decision theory, the *Condorcet jury theorem*, is that the majority judgement of the collective will be correct with increasing probability as the size of the group grows, under the weak assumption that every agent is mildly competent, that is, is correct with a probability greater than a half.[85] There are a number of ramifications of this basic result. For

---

[81] McPherson et al. (2001).

[82] DeGroot (1974) is an early model.

[83] See Easley and Kleinberg (2010), Chap. 16. On the general importance of initial stages in the process of judgement formation, see Arthur (1994).

[84] Molander (1986).

[85] With probability one, as group size $n \to \infty$.

groups with varying competence levels, it is possible to improve the accuracy of the common judgement by weighing the votes of the agents in accordance with their a priori competence.[86] The accuracy thus achieved is higher than that of the most competent expert in the network.

In summary, much depends on the way learning and judgement formation is organised within the network. The common adage about the wisdom of the crowds is true only under important restrictions, the most important being that of independence between the agents.[87]

Networks in Markets
Markets that exhibit network effects, like learning processes, display ambiguous behaviour. If the propensity of purchasing a good or adopting a new habit depends not only on price and quality but also on the choices made by other agents in the network, more than one equilibrium between price and quantity will normally be possible.[88] Obviously, a large company is in a better position than a small one when it comes to influencing the collective of consumers to converge on the equilibrium that yields the highest profits. Similar conclusions have been derived on the basis of different analyses focusing on the group of producers rather than consumers.[89]

The network approach to market analysis typically requires work-intensive data collection and is therefore limited in scope, focusing for instance on a local fish market or apparel industry. There are nonetheless situations where the data situation is more favourable, such as in analyses of the World Wide Web.[90] The *world trade* has been analysed using a network approach.[91] The trade network displays a highly skewed degree distribution and a high degree of clustering. Assortativity is negative, indicating a hub-and-spokes structure. The degree distribution is strongly skewed; on closer inspection, it turns out that poor countries follow a random pattern in their trade relations, whereas the node attachments of richer countries is more preference-governed, as could be expected.

5.5 Norms and Inequality

Norms appear in a number of situations in human life that are in some way problematic. Sometimes they are linked to mainly individual decision problems, such as whether to yield to some temptation or not, for instance, or whether to consume or to save. In such situations, the norm prescribes the "virtuous" line of

---

[86] Nitzan and Paroush (1982).

[87] Surowiecki (2004).

[88] Easley and Kleinberg (2010), Chap. 17.

[89] Ghemawat (1990).

[90] Albert et al. (1999).

[91] Serrano and Boguñá (2003).

action, the underlying justification normally being that this alternative tends to benefit the individual in the long run.

Most norms pertain to situations where individuals or groups interact in a social context, however. The origin is the same as in the individual dilemma—a recurrent situation that is in some way problematic and in which a virtuous alternative is beneficial to the collective. The role of the norm is to make the collectively rational outcome more likely by imposing a moral cost on the non-virtuous alternative.[92] We have encountered such dilemmas in all of the archetypical games. In the coordination game, a norm or convention serves the purpose of making the parties involved choose the same alternative. In prisoners' dilemma games, the norm prescribes the collectively rational alternative of cooperation—not to lie, not to steal, and so on. In games of conflict, the purpose of the norm will typically be not to cause more harm to the opponent than necessary, as unrestrained aggression may harm both parties and be dysfunctional from a social point of view.

How cooperation or norms can emerge in a population of selfishly acting agents is a long-standing problem both in biology and the social sciences. Cooperation means that some of the individual potential is forgone in order to help the group. But in the perspective of natural selection or competition, this represents a cost and therefore requires some specific mechanism that compensates for this cost. This problem is now relatively well understood, on the basis of such mechanisms as kin selection, direct reciprocity, indirect reciprocity, network reciprocity and group selection.[93] For each of these mechanisms, relatively simple rules specify whether natural selection can lead to cooperation.

A general conclusion from these theoretical analyses is that group size is important. The larger the group, the higher is the level of anonymity, the more complex is society and the more difficult it becomes to maintain cooperative alternatives with the aid of mechanisms founded on enlarged self-interest. This is where norms come in, as extra weight that may tilt the balance in favour of the collectively desirable alternative. In order to be efficient, norm systems have been developed and codified, and they have often been supported by religious infrastructures. Historically, there exists an association between social complexity and moralising gods, and such gods follow increases in social complexity.[94] Moralising gods and supernatural punishment for asocial acts, building on existing religious rituals, tend to appear only after the emergence of large societies.

In principle, norm systems may be symmetric in the sense that they benefit everyone in society on equal terms, by inducing people not to steal, not to hurt others, etcetera. In practice, however, the process of systematising and codifying norms has seldom been anchored in a democratic process. As a consequence, it reflects the unequal division of power in society, and the resulting norm system tends to reinforce existing, often starkly unequal, relationships.

---

[92]Ullmann-Margalit (1977).

[93]Nowak (2006).

[94]Whitehouse et al. (2019).

An illustrative example is the so-called "Noah's curse of Ham" in the *Book of Genesis*, according to which Noah condemns Ham's son Canaan to a life in slavery. In the *Book of Jubilees*, it is described how the world was divided among Noah's sons and Ham received Africa. In this way, the incident came to serve as an excuse for slavery in societies dominated by Christianity, for instance the modern United States.[95] During the Middle Ages in Europe, Ham was instead associated with the third estate, so Noah's curse was used as a divine justification for the oppression of the majority by the nobility and the priesthood.[96]

This form of religiously based, asymmetric norm systems is by no means limited to Christianity. Similar examples may be found, for instance, in Hinduism and Islam.

The driving force in the development sketched above is the increasing size of society. Not only size but also complexity, more specifically division of labour, has been historically important. The possibility of distinguishing one group from another is central. It is this distinction that divides society into two or more groups and thereby destroys the symmetric equilibria and paves the way for social equilibria that are asymmetric with respect to resources and power.[97] The resulting asymmetric equilibria may be less efficient than symmetric, egalitarian equilibria and yet persist for a long time. They may also be supported by linguistic conventions.[98]

5.6 Market Economies

In Sect. 1 above, the problem of inequality was reduced to a simple binary opposition between landlords and tenants or industrialists and labour. The perspective will now be widened to fully fledged market-based economies.

Remember that income and wealth inequality in the most advanced countries grew during the early phases of industrialisation and that there was a peak around the beginning of the twentieth century.[99] After that, inequality was reduced in most industrialised countries until around 1980, when a new period of growing inequality started. A noteworthy fact is that the relatively homogeneous group of OECD countries exhibits large differences in inequality. This raises a question about the long-term determinants of inequality.

A number of potentially relevant factors come to mind—economic, social, as well as political. In the first category, GDP, GDP growth, saving, financial markets, trade policy and public-sector policies are obvious candidates. Analysing these factors one

---

[95] Haynes (2002).

[96] Freedman (1999), Chap. 4.

[97] Hwang et al. (2016), Cochran and O'Connor (2019). See further Sect. 6.2, The Emergence of Classes.

[98] Naidu et al. (2017).

[99] For general overviews, see Alvaredo et al. (2017) and Roine and Waldenström (2015). Some of the main primary sources of the latter are Lindert (2000), Morrison (2000) and Roine et al. (2009).

by one yields insights, but it must be borne in mind that there are complex causal links between them internally and with external factors.

In a very basic way, the *GDP level* determines the span for inequality. In a population living at or close to the existence minimum, there is very little room for inequality. Therefore, inequality can be expected to grow with the GDP per capita. In more developed societies, the generally acceptable standard of living is above the existence minimum, however, and the politically available inequality span will in most cases be more restricted.[100] As a consequence, there is no simple relationship between GDP and inequality.

Simon Kuznets advanced the hypothesis that economic *growth dynamics* rather than GDP levels should determine the level of inequality.[101] Observing that wealthy persons save more, which should lead to permanently increasing inequalities in the long run, he sought an explanation for the rise and decline in inequality in the flow between sectors with different wage dispersions: "... the basic factor militating against the rise in upper income shares that would be produced by the cumulative effects of concentration of savings, is the dynamism of a growing and free economic society."[102] More precisely, a growing new sector in the economy would initially be subject to a labour shortage, which would force wages upwards and increase wage dispersion. As the supply of labour would eventually catch up with demands, dispersion would be reduced. Although the argument is logical, expanding sectors are normally too limited for this effect to provide an explanation for the twentieth-century movements of inequality.[103]

In a broad survey covering 26 countries over the period 1870–2010, the GDP *growth rate* turns out to have had a significant and positive effect on inequality.[104] The top percentile group benefits more than proportionately from high growth rates, which can be expected given its high dependence on capital income. The top decile group (excluding the top percentile), by contrast, is affected in the negative direction, because this group comprises highly salaried persons whose development is not directly tied to the general economic development. The effect on other decile groups is mixed.

As regards *savings*, Kuznets, as noted above, took it for granted that wealthy households will save more and that this propensity is a driving force in the direction of increased inequality. The question about the relationship between income and saving has turned out to be more intricate than he anticipated. That saving is positively related to current income is reasonable; a large fraction of household

---

[100] See Milanovic (2013) and Sect. 6.2 for a discussion in relation to social conflict.

[101] Kuznets (1955).

[102] Ibid., p. 11.

[103] For critical discussions of Kuznetsian models of explanation, see Lindert (2000) and Wisman (2017). The term *Kuznetsian* nonetheless has come to stay, it seems, and is used also for processes that are not Kuznetsian by nature; see, for example, Morrisson and Murtin (2013) and Milanovic (2016).

[104] Roine and Waldenström (2015), Section 6, for this and the following results unless otherwise stated. Data focus on the share of top incomes as a general indicator of inequality.

expenditure is bound by various commitments, so a current drop in income would lead to lower saving, and a temporary surplus would be saved for possible later negative turns. The fact that the distribution of consumption is closer to lognormal than income, which has a fatter tail, is an indication at the aggregate level.[105] The relationship between lifetime income and savings ratio is more open. A standard assumption in economic analysis has been that household savings are proportional to lifetime income, the alternative hypothesis being that wealthy households save more. If the latter hypothesis is correct, saving behaviour contributes to the concentration of wealth, as remaining capital is passed on to the following generation when a person dies. The answer is important to tax policies, because taxes on saving or consumption will differ in incidence if the latter hypothesis is true. Analysis of data from the United States shows that this is also the case. Evidence indicates that the driving forces behind this behavioural pattern are not differences in preferences but precautionary saving and the bequest motive, in the case of surplus.[106]

Financial development, measured as total bank deposits over GDP, market value of listed stocks and corporate bonds over GDP, or the sum of these two, tends to increase inequality. The effect is large within the high-income group (top decile group), but the overall effect is more limited.[107] The effect has to some extent been corroborated in studies covering a large number of countries. In developing countries, the development of financial markets is positive for growth and equality up to a certain point, where after further deepening of the financial market increases inequality.[108] Importantly, the outcome is conditioned on the quality of political institutions.[109]

There is little evidence of effects of *trade openness* on top income shares in the long run, which may go against expectations.[110] The explanation may lie in the political sphere. Countries that have opted for a free-trade-oriented regime have often developed a system of collective insurance—social security—against the consequences of structural change that may be brought about by the exposure to world market forces. In concrete terms, open economies have larger public sectors.[111] In this way, the benefits of international exchange are spread over a large part of the population, countervailing the direct effect that might benefit the wealthiest strata more than proportionally. This is a special case of a more general function of the welfare state, that is, as an insurance system pooling cost and benefits of

---

[105] Battistin et al. (2009).

[106] Dynan et al. (2004).

[107] Roine and Waldenström (2015).

[108] King and Levine (1993), Beck et al. (2007), de Haan and Sturm (2017) (121 countries during the period 1975–2005), Jauch and Watzka (2016) (130 countries 1960–2008), Brei et al. (2018) (97 counties 1989–2012).

[109] de Haan and Sturm (2017), Demirgüc-Kunt (2006).

[110] Roine and Waldenström (2015).

[111] Rodrik (1998), Bertola and Lo Prete (2008).

economic change and disruptions.[112] This is consistent with the observation that government spending seems associated with a larger income share for the nine lowest decile groups.

Top marginal tax rates have a significant effect on top income shares. This is a stable observation from different countries.[113] At least part of this effect can be ascribed to the possibilities of the highest income strata choosing different channels of remuneration.

5.7 Summary

The strategies that human beings apply when interacting with each other are answers to recurrent problems of learning and decision-making in nature and society. Observations on a high level of generalisation include the following:

- In many situations, highly complex interaction patterns can be reduced to binary exchange processes—between landowners and tenants, capital and labour, majority and minority, males and females, and so on. The outcome of these exchanges depends strongly on the bargaining strength of the two interacting parties—alternatives at hand, fallback positions, and ability to wait. Legislation and institutions affect the balance of power.
- Categories such as gender or ethnic background often cut across stratification according to incomes or assets, but correlations are strong. Visible characteristics are potential triggers for discrimination, implying that individuals are not remunerated according to their abilities.
- What is an efficient strategy in a given situation is not determined solely by the material conditions of the interacting parties but also by behaviour, which may be fixed in personal traits. Aggression is an important example of a behavioural strategy that may confer benefits on an otherwise mediocre agent in the social arena.
- In order to cope with the innumerable learning and decision problems in everyday life, human beings resort to thumb rules, which can often be perceived as hybrids between the ideal of fully rational decision-making and various imitation strategies. The typical outcome when such strategies are applied across a population of interacting individuals—whether in learning or decision-making—is a multiplicity of equilibria. Normally, the outcome will depend on prehistory, norms and other factors external to the decision situation in focus, and the well-being of the parties involved depends on which equilibrium materialises.

---

[112] Sinn (1995).

[113] Saez (2004) (USA), Saez and Veall (2005) (Canada), Roine and Waldenström (2008) (Sweden), Jäntti et al. (2009) (Finland), Atkinson and Leigh (2013) (Anglo-Saxon countries), Piketty et al. (2013) (18 OECD countries) and Rubolino and Waldenström (2020) (30 countries from the World Wealth and Income Database).

- In fully developed market economies, a number of factors work in the direction of increasing inequality. Wealthy households tend to save more, and empirically, both GDP growth and financial market development have tended to benefit upper income strata relatively. Network structures tend to increase inequality. By contrast, neither GDP levels nor trade openness have clear-cut effects on the distribution of income or wealth. Kuznetsian mechanisms of equalisation have limited power of explanation.

Bibliography

Abbott, D. H., et al. (2003). Are subordinates always stressed? A comparative analysis of rank differences in cortisol levels among primates. *Hormones and Behavior, 43*, 67–82.

Akenson, D. H. (2011). *Ireland, Sweden and the Great European Migration, 1815–1914*. Liverpool University Press.

Albert, R., et al. (1999). Diameter of the World Wide Web. *Nature, 401*, 130131.

Alfani, G. (2021). Economic inequality in preindustrial times: Europe and beyond. *Journal of Economic Literature, 59*(1), 3–44.

Altonji, J. G., & Blank, R. M. (1999). Race and gender in the labor market. Ch. 48 in O. Ashenfelter, & D. Card (Eds.), *Handbook of labor economics*, Vol. 3. Amsterdam: Elsevier.

Alvaredo, F., et al. (2017). *World inequality report 2018* (wir2018.wid.world).

Angrist, J. D. (2014). The perils of peer effects. *Labour Economics, 30*, 98–108.

Arrow, K. J. (1998). What has economics to say about racial discrimination? *Journal of Economic Perspectives, 12*(2), 91–100.

Arthur, W. B. (1994). *Increasing returns and path dependence in the economy*. The University of Michigan Press.

Asch, S. E. (1956). Studies of independence and conformity: I. A minority of one against a unanimous majority. *Psychological Monographs: General and Applied, 70*(9), 1–70.

Atkinson, A. B., & Leigh, A. (2013). The distribution of top incomes in five Anglo-Saxon countries over the long run. *Economic Record, 89*, Supplement S1, 31–47.

Atkinson, A. B., et al. (2011). Top incomes in the long run of history. *Journal of Economic Perspectives, 49*(1), 3–71.

Autor, D., et al. (2017). Concentrating on the fall of the labor share. *American Economic Review: Papers & Proceedings, 107*(5), 180–185.

Baldwin, R. (2016). *The great convergence. Information technology and the new globalization*. Belknap Press.

Baldwin, R. (2019). *The globotics upheaval. Globalization, robotics and the future of work*. Weidenfeld & Nicholson.

Ballester, C., et al. (2006). Who's who in networks. Wanted: The key player. *Econometrica, 74*(5), 1403–1417.

Bar Haim, E., & Shavit, Y. (2013). Expansion and inequality of educational opportunity: A comparative study. *Research in Social Stratification and Mobility, 31*, 22–31.

Barabási, A.-L. (2002). *Linked*. Perseus Publishing.

Bardhan, P., et al. (2000). Wealth inequality, wealth constraints, and economic performance. Ch. 10 in A. B. Atkinson & F. Bourguignon (Eds.), *Handbook of income distribution*, Vol. 1. Amsterdam: North-Holland.

Bassanini, A., & Manfredi, T. (2014). Capital's grabbing hand? A cross-industry analysis of the decline of the labor share in OECD countries. *Eurasian Business Review, 4*, 3–30.

Battistin, E., et al. (2009). Why is consumption more lognormal than income? Gibrat's law revisited. *Journal of Political Economy, 117*(6), 1140–1154.

Beck, T., et al. (2007). Finance, inequality and the poor. *Journal of Economic Growth, 12,* 27–49.

Bengtsson, E. (2014). Do unions redistribute income from capital to labour? Union density and wage shares since 1960. *Industrial Relations Journal, 45*(5), 389–408.

Bengtsson, E., & Waldenström, D. (2018). Capital shares and income inequality: Evidence from the long run. *The Journal of Economic History, 78*(3), 712–743.

Bertola, G., & Lo Prete, A. (2008). Openness, financial markets, and policies: Cross-country and dynamic patterns. *Annales d'Économie et Statistique, 95,* 167–182.

Bikhchandani, S., & Sharma, S. (2000). Herd behavior in financial markets. *IMF Economic Review, 47,* 279–310.

Black, D., et al. (2006). Why do minority men earn less? A study of wage differentials among the highly educated. *The Review of Economics and Statistics, 88*(1), 300–313.

Bloch, F., et al. (2008). Informal insurance in social networks. *Journal of Economic Theory, 143*(1), 36–58.

Boguslaw, J. (2017). When the kids are not alright. *Essays on childhood disadvantage and its consequences.* Ph.D. diss., Dept. of Economics, University of Stockholm.

Boissevain, J. (1974). *Friends of friends. Networks, manipulators and coalitions.* Basil Blackwell.

Boucher, V., et al. (2014). Do peers affect student achievement? Evidence from Canada using group size variation. *Journal of Applied Econometrics, 29*(1), 91–109.

Brei, M., et al. (2018). Financial structure and income inequality. *CEPR discussion paper* no. DP13330. : Centre for Economic Policy Research.

Buyukkececi, Z., et al. (2020). Family, firms, and fertility: A study of social interaction effects. *Demography, 57*(1), 243–266.

Camerlink, I., et al. (2015). Aggressiveness as a component of fighting ability in pigs using a game-theoretical framework. *Animal Behaviour, 108,* 183–191.

Carlsson, F., et al. (2007). Do you enjoy having more than others? Survey evidence of positional goods. *Economica, 74*(296), 586–598.

Chang, E. C., et al. (2000). An examination of herd behavior in equity markets: An international perspective. *Journal of Banking & Finance, 24,* 1651–1679.

Chetty, R., et al. (2014). Where is the land of opportunity? The geography of intergenerational mobility in the United States. *The Quarterly Journal of Economics, 129*(4), 1553–1623.

Christakis, N. A., & Fowler, J. H. (2008). The collective dynamics of smoking in a large social network. *New England Journal of Medicine, 358*(21), 2249–2258.

Claessens, S., et al. (2002). Disentangling the incentive and entrenchment effects of large shareholdings. *Journal of Finance, LVII*(6), 2741–2771.

Cochran, C., & O'Connor, C. (2019). Inequality and inequity in the emergence of conventions. *Politics, Philosophy & Economics, 18*(3), 264–281.

Cowlishaw, G., & Dunbar, R. I. M. (1991). Dominance rank and mating success in male primates. *Animal Behaviour, 41*(6), 1045–1056.

Davies, J. B., & Shorrocks, A. F. (2000). The distribution of wealth. Ch. 11 in A. F. Atkinson, & F. Bourguignon (Eds.), *Handbook of income distribution,* Vol. 1, 605–675.

de Haan, J., & Sturm, J.-E. (2017). Finance and income inequality: A review and new evidence. *European Journal of Political Economy, 50,* 171–195.

DeGroot, M. H. (1974). Reaching a consensus. *Journal of the American Statistical Association, 69,* 118–121.

Deininger, K., & Olinto, P. (2000). Asset distribution, inequality, and growth. *Policy research working paper* 2375. Washington, DC: The World Bank.

Demirgüç-Kunt, A. (2006). Finance and economic development: Policy choices for developing countries. *World bank policy research working paper* 3955. Washington, DC: The World Bank.

Dong, Y., et al. (2017). Shifting diets and the rise of male-biased inequality on the Central Plains of China during Eastern Zhou. *Proceedings of the National Academy of Sciences, 114*(5), 932–937.

Dynan, K. E., et al. (2004). Do the rich save more? *Journal of Political Economy, 112*(2), 397–444.

Easley, D., & Kleinberg, J. (2010). *Networks, crowds and markets.* Cambridge University Press.

Elborgh-Woytek, K., et al. (2013). Women, work, and the economy: Macroeconomic gains from gender equity. *IMF Staff discussion note* 13/10. Washington, DC: The International Monetary Fund.

Faccio, M., & Lang, L. H. P. (2002). The ultimate ownership of Western European corporations. *Journal of Financial Economics, 65*, 365–395.

Favati, A., et al. (2017). Individual aggression, but not winner–loser effects, predicts social rank in male domestic fowl. *Behavioral Ecology, 28*(3), 874–882.

Freedman, P. (1999). *Images of the medieval peasant.* Stanford University Press.

Friedman, M. (1962). *Capitalism and freedom.* Chicago University Press.

Ghemawat, P. (1990). The snowball effect. *International Journal of Industrial Organization, 8*, 335–351.

Gigerenzer, G., & Todd, P. M. (1999). *Simple heuristics that make us smart.* Oxford University Press.

Granovetter, M. S. (1973). The strength of weak ties. *American Journal of Sociology, 78*(6), 1360–1380.

Gustavsson, M., & Jordahl, H. (2008). Inequality and trust in Sweden: Some inequalities are more harmful than others. *Journal of Public Economics, 92*, 348–365.

Hajnal, J. (1982). Two kinds of preindustrial household formation system. *Population and Development Review, 8*(3), 449–494.

Hamilton, K., et al. (2017). Social capital, trust, and well-being in the evaluation of wealth. Chapter 12 in K. Hamilton, & C. Hepburn (Eds.), *National wealth. What is missing, and why it matters.* Oxford: Oxford University Press.

Hanauske-Abel, H. M. (1996). Not a slippery slope or sudden subversion: German medicine and National Socialism in 1933. *British Medical Journal, 313*(7070), 1453–1463.

Haynes, S. R. (2002). *Noah's curse. The biblical justification of American slavery.* Oxford University Press.

Hensvik, L., & Nilsson, P. (2010). Businesses, buddies and babies: Social ties and fertility at work. *IFAU working paper* 2010:9, Uppsala.

Hirsch, F. (1976). *Social limits to growth.* Harvard University Press.

Hoff, K. (2016). Caste system. *World bank policy research working paper* 7929. Washington, DC: The World Bank.

Holekamp, K. E., & Strauss, E. D. (2016). Aggression and dominance: An interdisciplinary overview. *Current Opinion in Behavioral Sciences, 12*, 44–51.

Höllinger, F., & Haller, M. (1990). Kinship and social networks in modern societies: A cross-cultural comparison among seven nations. *European Sociological Review, 6*(2), 103–124.

Hwang, S.-H., et al. (2016). Social conflict and the evolution of unequal conventions. *Working paper*, Santa Fe Institute, CA.

IMF. (2017). *World economic outlook April 2017.* International Monetary Fund.

Ings, T. C., et al. (2009). Ecological networks: Beyond food webs. *Journal of Animal Ecology, 78*, 253–269.

Ireland, P. (2005). Shareholder primacy and the distribution of wealth. *Modern Law Review, 68*(1), 49–81.

Jackson, M. O. (2008). *Social and economic networks.* Princeton University Press.

Janis, I. (1982). *Groupthink* (2nd ed.). Houghton Mifflin.

Jäntti, M., et al. (2009). Trends in top income shares in Finland. In A. B. Atkinson & T. Piketty (Eds.), *Top incomes: A global perspective* (Vol. II). Oxford University Press.

Jauch, S., & Watzka, S. (2016). Financial development and income inequality: A panel data approach. *Empirical Economics, 51*, 291–314.

Joerges, C., & Ghaleigh, N. S. (Eds.). (2003). *Darker legacies of law in Europe. The Shadow of National Socialism and Fascism over Europe and its Legal Traditions.* Hart Publishing.

Jordà, O., et al. (2019). The rate of return on everything, 1870–2015. *The Quarterly Journal of Economics, 134*(3), 1225–1298.

Karabarbounis, L., & Neiman, B. (2014). The global decline of the labor share. *The Quarterly Journal of Economics, 129*(1), 61–103.

Karadja, M., & Prawitz, E. (2019). Exit, voice, and political change: Evidence from Swedish mass migration to the United States. *Journal of Political Economy, 127*(4), 1864–1925.

Kehrig, M., & Vincent, N. (2017). Growing productivity without growing wages: The micro-level anatomy of the aggregate labor share decline. *CESifo working paper* No. 6454, Munich.

King, R. G., & Levine, R. (1993). Finance and growth: Schumpeter might be right. *The Quarterly Journal of Economics, 108*(3), 717–737.

Kirman, A. P. (1983). Communication in markets. A suggested approach. *Economics Letters, 12*, 101–108.

Kirman, A. P., & Vriend, N. J. (2000). Learning to be loyal. A study of the Marseille fish market. In D. Delli Gatti et al. (Eds.), *Interaction and market structure: Essays on heterogeneity in economics*. Springer.

Kramer, M. R., & Hogue, C. R. (2009). Is segregation bad for your health? *Epidemiologic Reviews, 31*, 178–194.

Krueger, A. B. (1999). Measuring labor's share. *American Economic Review, 89*, 45–51.

Krusell, P., & Smith, A. A., Jr. (2015). Is Piketty's 'second law of capitalism' fundamental? *Journal of Political Economy, 123*(4), 725–748.

Kuznets, S. (1955). Economic growth and income inequality. *American Economic Review, XLV*(I), 1–28.

La Porta, R., et al. (1999). Corporate ownership around the world. *The Journal Of Finance, LIV*(2), 471–517.

Liljeros, F., et al. (2001). The web of human sexual contacts. *Nature, 411*(21 June), 907–908.

Lindert, P. H. (2000). Three centuries of inequality in Britain and America. Ch. 3 in A. B. Atkinson, & F. Bourguignon (Eds.), *Handbook of income distribution*, Vol. 1. Amsterdam: North-Holland.

Lipton, M. (2009). *Land reform in developing countries. Property rights and property wrongs.* Routledge.

Massey, D. S., et al. (2009). The Changing Bases of Segregation in the United States. *Annals of the American Academy of Political and Social Science, 626*(1), 74–90.

McEvoy, J., et al. (2013). The role of size and aggression in intrasexual male competition in a social lizard species, *Egernia whitii. Behavioral Ecology and Sociobiology, 67*, 79–90.

McPherson, M., et al. (2001). Birds of a feather: Homophily in social networks. *Annual Review of Sociology, 27*, 415–444.

Milanovic, B. (2013). The inequality possibility frontier. Extensions and new applications. *Policy research working paper* 6449. Washington, DC: The World Bank.

Milanovic, B. (2016). *Global inequality. A new approach for the age of globalization.* Belknap Press.

Mitchell, W. P. (1973). The hydraulic hypothesis: A reappraisal. *Current Anthropology, 14*(5), 532–534.

Molander, P. (1986). Induction of categories: The problem of multiple equilibria. *Journal of Mathematical Psychology, 30*(1), 42–54.

Morrison, C. (2000). Historical perspectives on income distribution: The case of Europe. Ch. 4 in A. B. Atkinson & F. Bourguignon (Eds.), *Handbook of income distribution*, Vol. 1. Amsterdam: North-Holland.

Morrisson, C., & Murtin, F. (2013). The Kuznets curve of human capital inequality: 1870–2010. *Journal of Economic Inequality, 11*, 283–301.

Naidu, S., et al. (2017). The evolution of egalitarian sociolinguistic conventions. *American Economic Review: Papers & Proceedings, 107*(5), 572–577.

Newman, M. E. J. (2003). The structure and function of complex networks. *SIAM Review, 45*(2), 167–256.

Nitzan, S., & Paroush, J. (1982). Optimal decision rules in uncertain dichotomous choice situations. *Theory and Decision, 13*, 129–138.

Nowak, M. A. (2006). Five rules for the evolution of cooperation. *Science, 314*(5805), 1560–1563.

Ostrom, E. (1992). *Crafting institutions for self-governing irrigation systems*. Institute for Contemporary Studies.

Pichler, F., & Wallace, C. (2009). Social capital and social class in Europe: The role of social networks in social stratification. *European Sociological Review, 25*(3), 319–332.

Piketty, T., et al. (2013). Optimal taxation of top incomes: A tale of three elasticities. *American Economic Journal: Economic Policy, 6*(1), 230–271.

Pross, C. (1991). Breaking through the postwar coverup of Nazi doctors in Germany. *Journal of Medical Ethics, 17*(Supplement), 13–16.

Provençal, N., et al. (2015). The developmental origins of chronic physical aggression: Biological pathways triggered by early life adversity. *The Journal of Experimental Biology, 2015*(218), 123–133.

Putnam, R. (2001). Social capital: Measurement and consequences. *Canadian Journal of Policy Research, 2*, 41–51.

Ricardo, D. (1821). *The principles of taxation and political economy*. J. M. Dent.

Rodrik, D. (1998). Why do more open economies have bigger governments? *Journal of Political Economy, 106*(5), 997–1032.

Rodrik, D. (1999). Democracies pay higher wages. *The Quarterly Journal of Economics, CXIV*(3), 707–738.

Roine, J., & Waldenström, D. (2008). The evolution of top incomes in an egalitarian society: Sweden, 1903–2004. *Journal of Public Economics, 92*, 366–387.

Roine, J., & Waldenström, D. (2015). Long-run trends in the distribution of income and wealth. Ch. 7 in Edited by A.B. Atkinson, & F. Bourguignon (Eds.), *Handbook of income distribution*, Vol. 2, pp. 469–592.

Roine, J., et al. (2009). The long-run determinants of inequality: What can we learn from top income data? *Journal of Public Economics, 93*(7–8), 974–988.

Rubolino, E., & Waldenström, D. (2020). Tax progressivity and top incomes. Evidence from tax reforms. *The Journal of Economic Inequality, 18*, 261–289.

Sacerdote, B. (2011). Peer effects in education: How might they work, how big are they and how much do we know thus far? Ch. 4 in E. A. Hanushek et al. (Eds.), *Handbook of the economics of education*, Vol. 3, pp. 249–277.

Saez, E. (2004). Reported incomes and marginal tax rates, 1960–2000: Evidence and policy implications. In J. Poterba (Ed.), *Tax policy and the economy* (Vol. 18). MIT Press.

Saez, E., & Veall, M. R. (2005). The evolution of high incomes in Northern America: Lessons from Canadian evidence. *American Economic Review, 95*, 831–849.

Sapolsky, R. M. (2004). Social status and health in humans and other animals. *Annual Review of Anthropology, 33*, 393–418.

Scarr, C. (Ed.). (2018). *The human past. World prehistory and the development of human societies*. Thames & Hudson.

Scheidel, W. (2017). *The great leveler. Violence and the history of inequality from the stone age to the twenty-first century*. Princeton University Press.

Serrano, M. A., & Boguñá, M. (2003). Topology of the world trade web. *Physical Review E, 68*, 015101(R).

Shenk, M. K., et al. (2010). Intergenerational wealth transmission among agriculturalists foundations of agrarian inequality. *Current Anthropology, 51*(1), 65–83.

Sherif, M. (1936). *The psychology of social norms*. Harper Collins.

Sinn, H.-W. (1995). A theory of the welfare state. *Scandinavian Journal of Economics, 97*(4), 495–526.

Smith, A. (1776). *An inquiry into the nature and causes of the wealth of nations*. Several later editions.

Stolleis, M. (1994). *Studien zur Rechtsgeschichte des Nationalsozialismus*. Frankfurt am Main: Suhrkamp Verlag. Eng. trans. *The Law under the Swastika. Studies on Legal History in Nazi Germany*. Chicago: University of Chicago Press (1998).

Stotsky, J. G., et al. (2016). Trends in gender equality and women's advancement. *IMF WP/16/21*. Washington, DC: International Monetary Fund.

Strang, D., & Soule, S. A. (1998). Diffusion in organizations and social movements: From hybrid corn to poison pills. *Annual Review of Sociology, 24*, 265–290.

Surowiecki, J. (2004). *The wisdom of crowds*. Little, Brown.

Ullmann-Margalit, E. (1977). *The emergence of norms*. Clarendon Press.

Uslaner, E. (Ed.). (2018). *The Oxford handbook of social and political trust*. Oxford University Press.

Uzzi, B. (1996). The sources and consequences of embeddedness for the performance of organizations: The network effect. *American Sociological Review, 61*(4), 674–698.

Visser, J., & Checchi, D. (2009). Inequality and labor market: Unions. In W. Salverda et al. (Eds.), *The Oxford handbook of economic inequality*. Oxford University Press.

Whitehouse, H., et al. (2019). Complex societies precede moralizing gods throughout world history. *Nature, 20*(March 2019), 226–229.

Winson, G. C., & Zimmermann, D. J. (2014). Peer effects in higher education. In C. Hoxby (Ed.), *College choices: The economics of where to go, when to go, and how to pay for it*. University of Chicago Press.

Wisman, J. D. (2017). Politics, not economics, ultimately drives inequality. *Challenge, 60*(4), 347–367.

Wittfogel, K. (1957). *Oriental despotism. A comparative study of total power*. Yale University Press.

Wolf, M. et al. (2007). Life-history trade-offs favour the evolution of animal personalities. *Nature, 447*, 31 May 2007, 581–585.

World Bank. (2005). *Equity and development. World development report 2006*. The World Bank.

Chapter 6
Models of Interaction*

Interaction in small and large populations has been analysed using a variety of models that reflect the multi-faceted relationships between human beings—cooperation, competition, conflict, and mixtures of these. Needless to say, all these models are simplifications, sometimes brutal, and normally aim at illustrating one particular aspect of social interaction, or answering one particular question. Examples include "Under what conditions is cooperation between selfish individuals possible?" and "Which are the mechanisms that generate segregation in residential areas and schools?"

Against the backdrop of Ricardo's claim that the distribution of the surplus produced in a society is the principal problem in political economy, the basic question is perhaps what resource distribution we should expect in a given social structure. Is an egalitarian distribution stable? If not, what are the mechanisms of interaction behind increasing inequality? The succinct overview presented in the previous chapter indicates that the answer to the first of these questions is in the negative. The focus in the present chapter will therefore be on presenting models that explain how asymmetric patterns of distribution can emerge and stabilise.

6.1 Bilateral Negotiation

Exchange in various forms is basic to human interaction. Some social theorists have even perceived it as the equivalent of a chemical bond in society[1]—an elementary relationship from which complex social structures are built. Many of these exchange relationships presuppose bargaining in some form, explicit or implicit. In market-type situations, bargaining is more or less the defining operation, and markets have indeed penetrated very far into the social fabric of modern societies. But bargaining

---

[1] See Bredemeier (1978) for a survey of exchange theories.

© The Author(s), under exclusive license to Springer Nature Switzerland AG 2022
P. Molander, *The Origins of Inequality*,
https://doi.org/10.1007/978-3-030-93189-6_6

is not restricted to markets. It shows up in pre-market, even prehistoric, archetypical situations, and it has survived in petrified form in many fields as yet relatively untouched by markets.

We tend to think of bargaining as explicit haggling over prices and other conditions of exchange. The archetype of bargaining situations is that of a seller of a good and a potential buyer meeting in the classical boisterous marketplace or in an antique shop. The position of this archetype is of course reinforced by the use of the market as the dominant metaphor in economic theory and practice. Even if such situations have been common in history and remain so in many parts of the world, bargaining in markets tends to assume other forms in developed industrial countries. This is not necessarily the explicit form that we normally associate with bargaining in a marketplace.

Whatever information the two parties have at their disposal, they are likely to be able to predict quite soon what will be the approximate price of the good and so go for that price rather quickly. With some more information, they may know more or less certainly beforehand what the price will be. Even without prior information, the process may converge quickly with the aid of thumb rules. Splitting the difference is a common rule.

The endpoint of this process is to abolish the process of bargaining altogether. This is the normal state of affairs for everyday consumer goods in developed market economies. The sellers offer goods with price tags, and the consumers decide whether they are prepared to buy at the price demanded.

Non-market decisions in political or social life are similar. In some situations, such as in post-war negotiations over how to divide a territory or in EU council meetings, the formal procedure is important, if for no other reason than to show citizens at home that a serious effort has been made. But in less dramatic contexts, no formal negotiation is entered, because both parties know roughly what the outcome will be. Working life offers many examples where colleagues divide tasks without discussion, even though they may have somewhat different ideas on what should be expected from each and everyone.

In these examples, the parties involved make a rapid or even subconscious assessment of the situation and decide how to act—in effect, a sort of implicit bargaining. Even a short list of examples such as this suffices to show that implicit bargaining may be equally or more important than formal or explicit bargaining. The absence of a formal process does not indicate full harmony of interests. On the contrary, beneath the surface there may be severe conflicts, but because the parties expect a certain outcome, bargaining becomes superfluous.

As illustrated in the previous chapter, the basic structure of many important high-level exchange relations can be reduced to a two-party bargaining process—between landowners and tenants, between capital and labour, and so on. The paradigmatic model of this process and its outcome is the Nash bargaining model. Remember (Sect. 2.4) that the Nash solution can be derived in several different ways—on the basis of pre-specified, desired characteristics of the solution, as the outcome of an explicit bargaining procedure or as the product of an evolutionary process. There are consequently strong reasons for taking the Nash solution as the point of departure for an analysis of the stability of an egalitarian equilibrium.

Instability of the Egalitarian Equilibrium

The bargaining strength of a party to a negotiation is determined by several factors. The default value—what will happen if the negotiation fails—is obviously of paramount importance. It determines the ability to take risks and to wait if the party in question is bound to the scene of negotiation. There may also be exit options that could strengthen a party's position. In the case of bargaining between individuals, personality traits such as patience or perseverance may affect the outcome.

The important bargaining games that are in focus are repeated. This implies that the outcome of one bargaining session affects the initial value for the following session. If one party happens to gain somewhat more than the other, that will be an advantage in the session to follow, assuming that the surplus is at least partially carried over to the following period. This is an indication that a symmetric equilibrium may be unstable. This will not hold in general, however; a necessary and sufficient condition will be presented in what follows.

Define as usual a utility function $u(x)$, where x is the quantity at the person's disposal. The function $u(\cdot)$ is assumed to satisfy the standard conditions of being *strictly growing* ($u'(x) > 0$ for all $x > 0$) and *strictly concave* ($u''(x) < 0$ for all $x > 0$).

Another important parameter is *risk aversion*. As derived by Pratt and Arrow,[2] the absolute risk aversion (ARA) is related to the utility function $u(x)$ via.

$$\mathrm{ARA}(x) = -u''(x)/u'(x)$$

Intuitively, an agent that controls larger resources is prepared to take higher risks. This has been empirically confirmed. A broad survey of the literature on risk aversion concludes that the assumption on decreasing absolute risk aversion has very strong support.[3]

Assume now that there are two parties, *1* and *2*, whose assets are designated by x_1 and x_2, respectively, and that bargaining outcomes are determined by Nash's bargaining theorem. Further:

1. The two parties have the same utility function $u(x_i)$, $i = 1, 2$. This function satisfies the standard assumptions of strict growth and strict concavity, $u'(x_i) > 0$, $u''(x_i) < 0$.
2. Absolute risk aversion $|u''(x_i)/u'(x_i)|$ is strictly decreasing in x_i.
3. The two parties bargain over a quantity, which may be set to 1 without loss of generality. Party *1* receives x and party *2* receives $(1 - x)$.

Proposition Under assumptions 1–3, the egalitarian equilibrium $x_1 = x_2$ is unstable, that is, perturbations from this equilibrium will be magnified. Condition 2 on decreasing risk aversion is necessary.

---

[2] Pratt (1964), Arrow (1965).
[3] Meyer and Meyer (2006), Sect. 3.1.

Proof

Sufficiency: Assume without loss of generality that the initial point is (0,0) and that $u(0) = 0$.

The Nash solution to the bargaining to the problem is the point (x_1, x_2) that maximises the product

$$f(x_1, x_2) = u(x_1) \cdot u(x_2) = u(x) \cdot u(1 - x)$$

Suppose now that the original equilibrium is disturbed, so that party *1* starts at *A*, where *A* is a positive quantity. The function to be maximised now becomes

$$g_A(x) = [u(x + A) - u(A)] \cdot u(1 - x)]$$

The optimum is given by the (unique) root of the equation

$$u'(x + A) \cdot u(1 - x) - [u(x + A) - u(A)] \cdot u'(1 - x) = 0.$$

Alternatively,

$$[u(x + A) - u(A)]/u'(x + A) = u(1 - x)/u'(1 - x).$$

The left-hand side is strictly increasing in *x*, while the right-hand side is strictly decreasing, confirming that there exists a unique root to the equation. Showing that the part allotted to party *1* through bargaining increases if the original wealth of party *1* is greater than that of party *2* is equivalent to showing that the left-hand side function $h_x(A)$ defined by

$$h_x(A) = [u(x + A) - u(A)]/u'(x + A)$$

decreases with *A*. Consider the derivative with respect to *A*:

$$\partial_A h_x(A) = \{u'(x + A) \cdot [u'(x + A) - u'(A)] - [u(x + A) - u(A)] \cdot u''(x + A)\}/\{u'(x + A)\}^2$$

We can write

$$u'(x + A) - u'(A) = \int u''(t)\, dt$$

$$u(x + A) - u(A) = \int u'(t)\, dt$$

where in both cases the integral is computed from *(A)* to *(x + A)*. Consequently,

$$\partial_A h_x(A) = \frac{\{u'(x+A) \cdot \int u''(t)dt - \int u'(t)dt \cdot u''(x+A)\}}{\{u'(x+A)\}^2}$$

$$= \frac{\{u'(x+A) \cdot \int u'(t) \cdot [u''(t)/u'(t)]dt - \int u'(t)dt \cdot u''(x+A)\}}{\{u'(x+A)\}^2}$$

$$< \{u'(x+A) \cdot max\ [u''(t)/u'(t)] - u''(x+A)\} \int u'(t)dt/\{u'(x+A)\}^2$$

where the maximum is taken over the interval $[(A), (x + A)]$. Given that $|u''(x)/u'(x)|$ is strictly decreasing in x, that maximum value is assumed at the right end point of the interval and equals $u''(x + A)/u'(x + A)$. This makes the expression in parentheses before the integral equal to 0. In consequence, the function $h_x(A)$ is strictly decreasing in A, as required.

Necessity: A counterexample is sufficient to prove the necessity of decreasing absolute risk aversion. Given A, assume that the negotiation is about the quantity A and consider the utility function

$$u(x) = \begin{cases} x, & 0 \leq x \leq 1.2\,A \\ 1.2\,A + 0.2 \cdot (x - 1.2\,A), & x > 1.2\,A \end{cases}$$

This function has a discontinuity in the first derivative at $x = 1.2\,A$, and the second derivative does not exist there, but the function can be regularised in an arbitrarily small neighbourhood around this point without altering the conclusion.

As before, let the two parties numbered *1* and *2* have this basic utility function, the difference being that party *1* starts at A, yielding the utility function

$$u_1(x) = \begin{cases} x, & 0 \leq x \leq 0.2\,A \\ 0.2\,A + 0.2 \cdot (x - 0.2\,A), & x > 0.2\,A \end{cases}$$

In utility space, the budget restriction $x_1 + x_2 \leq A$ takes the form

$$u_2(u_1) = \begin{cases} A - u_1 & 0 \leq u_1 \leq 0.2\,A \\ 0.8\,A - 5 \cdot (u_1 - 0.2\,A) & 0.2\,A < u_1 \leq 0.36\,A \end{cases}$$

It is easily verified that the Nash solution to this negotiation is at the discontinuity, yielding $x_1 = 0.2\,A$, $x_2 = 0.8\,A$. Party 2 thus gets a larger share, although party *1* starts with larger assets. ■

The above result is based on a set of highly simplifying assumptions. The negotiation process involves only two parties, which for the purpose of analysis can be considered as single decision-makers. No attempt is made to model the processes in which each party reaches a conclusion about the strategy to play.

Further, the parties are assumed to be myopic. They may be aware of a continuation of the exchange but do not exploit this possibility, at least not in the medium term. In order to include this strategic possibility, it would be necessary to value future risks versus future gains, which requires a discounting function. The standard choice of a constant discount factor has weak empirical support, and the introduction of a time-varying discount factor will give rise to qualitatively new behaviour.[4] In an evolutionary setting, strategies may be modified as the parties gather experience from the exchange process.

In spite of these simplifications, the result is suggestive. It indicates the presence of *symmetry breaking*, that is, the transition from an egalitarian, symmetric but unstable equilibrium into an asymmetric regime characterised by more or less pronounced inequality. This simple description is silent on the development following the break-down of the egalitarian equilibrium, but there is no mechanism in the bilateral negoti-ation structure itself that brings a process of increasing inequality to a halt. What sets the limit to asymmetry must consequently be some ad hoc borderline, defined by self-restraint on the part of the stronger party, a state of more or less pronounced conflict, or an exit alternative that permits the weaker party to leave the game.

6.2 Conflict

A natural follow-up to the previous section is an analysis of the process following the breakdown of the egalitarian equilibrium. Three examples will be given here. The first concerns the emergence of classes, with a more or less fixed division of roles between exploited and exploiters. The second links the degree of exploitation to the risk of open conflict. The third provides tools for the analysis of conflict itself, primarily the hawk-dove game.

The Emergence of Classes

Using a bargaining model that is a variant of the Nash model, Axtell, Epstein and Young have analysed the process by which discrimination and classes can emerge in a society.[5] As in the previous game, there is a quantity—a harvest, an annual value added—that is to be shared between two parties. Both parties suggest a partitioning, and if these are compatible, the quantity under negotiation is divided accordingly. If they are not, both parties get zero. There is no bargaining involved, either explicit or implicit, so accommodation takes place at the aggregate level, as successful strate-gies multiply and less successful ones are reduced in number or eliminated altogether.

In order to simplify analysis, it is assumed that there are three strategies available: low (L), medium (M) and high (H). The total quantity is 100, and the three demand alternatives open to the parties are 30, 50 and 70, respectively. This yields the game matrix in Fig. 6.1.

---

[4] See, for example, Vieille and Weibull (2008).

[5] Axtell et al. (2001).

Fig. 6.1 The
Demand Game

	L	M	H
L	30, 30	30, 50	*30, 70*
M	50, 30	*50, 50*	0,0
H	*70,30*	0,0	0,0

Three of the resulting combinations yield positive payoffs that are Nash equilibria, marked in italics along one of the diagonals. Any of these, or a mixture, is a possible outcome of the game. The analysis proceeds in two steps. In the first, a homogeneous population is assumed, in the second, a population in which it is possible to make the distinction between two categories of people.

Assume first that population is homogeneous. In each game round, two players are selected at random and make a choice based on their previous experience, stored in a memory of given length. At any moment in time, the set of individual memories defines the state of the society in question. Each individual makes a forecast of the other player's choice, which is based on the relative frequencies of L, M and H in previous meetings as far as the memory reaches. Someone who expects an M will choose M, whereas someone expecting H will play L, and vice versa. There is a stochastic element, in that the individual chooses at random between the three strategies available with probability ε, and the strategy that maximises the payoff given the predicted strategy of the other player with probability $(1 - \varepsilon)$. This random element can correspond to a pure error in the implementation of the best response, to an attempt to gain information, or simply to an interest in trying something different.

This set-up defines a Markov chain with well-defined transition probabilities. Put succinctly, there are two possible outcomes of this process, an egalitarian and a fractious. In the egalitarian regime, most of the players have met with M demands and consequently propose M themselves. A norm of equal division is established, characterised by a common expectation that everyone will play and the actual choice of M by an overwhelming majority. This is an efficient outcome, given that the average payoff is close to the maximal 50.

In the fractious outcome, players have mixed memories mainly consisting of Hs and Ls and so choose with similar probabilities between L and H. This outcome is far from efficient. About half of the time, players will choose differently and receive 50 on average, but half of the time they will either both choose L and receive 30 or both choose H and receive 0. This outcome hurts everybody equally on average, because players will move permanently between different memory states.

Which of these two outcomes is the more likely depends on initial conditions or prehistory. A detailed stochastic analysis shows that, assuming a sufficiently large population and sufficiently long memories, the egalitarian regime is more likely in the long run.[6] For reasonable memory sizes, the transition time to this regime may be long, however, so the long run may be very long—of the order of 10^6 units of time.

---

[6]This equilibrium is *stochastically stable* (Foster and Young 1990); see also Young (1998).

Turning now to the heterogeneous case, assume that there is some discernible sign—skin colour, language or religious affiliation marked by some visible tag—that makes it possible to divide the population into two disjoint categories, A and B. In this way, a two-population game is defined. The sign of distinction need have no link whatsoever to productivity or moral constitution, or to any other substantial characteristic, but is simply a tag. In this society, players differentiate their responses between players of their own group and players of the opposite group.

As in the homogeneous case, there is an egalitarian equilibrium, in which almost everybody opts for the medium alternative and expects everyone else to do the same. No difference is made between the two categories. This is also the main alternative for the long run, but as in the previous situation, there are other, less harmonious outcomes that may be long-lived.

There is first the possibility of equality between but not within types. This is basically a duplication of the previous fractious outcome in the two subpopulations, which we know to be inefficient. There are also outcomes where members of two groups are treated differently—members of group B behave aggressively towards members of group A, and the latter expect this behaviour and adapt their expectations accordingly. While the upper group, B, treats other members of the same group equally, there are two possible outcomes for members of group A. Either they treat each other equally, or they fall into the fractious regime, unable to unite against the exploiting group B. Whether one or the other of these alternatives will materialise again depends on prehistory.

Dynamic analysis of these latter regimes shows that transition times to the egalitarian regime are even longer than in the homogeneous case. Segregation and inequality may consequently persist for long periods.

The Risk of Conflict

Inequality requires a surplus. If a society produces precisely what is needed for survival, there is no room for inequality. If the level of production relative to population increases, so can inequality. For any level of GDP per capita, one can compute the maximal Gini coefficient G^* that is compatible with the restriction that the whole population receives the physiological minimum necessary for survival. The *inequality possibility frontier* is the locus of G^* as GDP increases from its minimum to its maximum value.[7]

In most societies, the Gini coefficient will be smaller than its maximum value. The *inequality extraction ratio* (IER) is defined as the ratio between the actual (measured) value G and the maximum value:

$$IER = G/G^*$$

---

[7] Milanovic et al. (2011).

If it is further assumed that society is controlled by a numerically small elite, and that the average income in society can be expressed as a multiple α (>1) of the physiological minimum, the IER can be expressed as[8]

$$IER = (\alpha - 1)/\alpha$$

Empirically, the extraction ratio showed a tendency to decrease in premodern societies when GDP per capita increased.[9] There are also interesting results on this ratio and its decomposition at the global level.[10] Here, the focus will be on the link between the extraction ratio and conflict.[11]

As a country leaves the level of absolute poverty, what is considered to be the minimum level of income can be expected to rise as well, although more slowly than average income. Adam Smith noted[12]:

> By necessaries I understand, not only the commodities which are indispensably necessary for the support of life, but whatever the custom of the country renders it indecent for creditable people, even of the lowest order, to be without. A linen shirt, for example, is, strictly speaking, not a necessary of life. The Greeks and Romans lived, I suppose, very comfortably, though they had no linen. But in the present times, through the greater part of Europe, a creditable day-labourer would be ashamed to appear in public without a linen shirt, the want of which would be supposed to denote that disgraceful degree of poverty, which, it is presumed, no body can well fall into without extreme bad conduct.

Modern estimates of the elasticity of the perceived minimum level with respect to income yield values in the vicinity of 0.5, that is, the perception of what is an acceptable minimum rises with income but more slowly than proportionally.[13] The fact that the perceived acceptable minimum increases with mean income in a society implies that the definition of the inequality extraction ratio should be based on this minimum rather than the physiological minimum. The scope for extraction by the elite is more limited than the previous definition would suggest. For instance, the extraction ratio for the United States around 1850 according to this definition is shifted from just over 0.6 to 0.85.[14]

When testing whether inequality is a source of conflict, Collier and Hoeffler relied on the Gini coefficient as a measure of inequality and came up with a negative answer.[15] As pointed out earlier (Sect. 2.2), the Gini coefficient is a blunt tool, and the negative outcome of the hypothesis test is not a final answer. Milanovic used the inequality extraction ratio with a socially defined minimum subsistence income in

---

[8] Ibid.

[9] Milanovic (2018).

[10] See Sect. 7.2.

[11] Milanovic (2013).

[12] Smith (1776), Book V, Part II.

[13] See references in Milanovic (2013).

[14] Milanovic (2013), Fig. 4.

[15] Collier and Hoeffler (2004).

order to test the hypothesis and found significant links between this measure and the intensity of civil conflict.[16] This is, of course, not the only factor of importance; ethnic and linguistic fractionalisation are strongly correlated with both duration of conflict and casualty rates. A more disaggregated measure of inequality in combination with the exploitation ratio perspective could possibly increase the explanatory power of the inequality factor.

The Dynamics of Conflict

Remember that, reduced to bare essentials, there are only three different categories of 2x2 symmetric games:

- *Dominant strategy games*: One pure strategy strictly dominates the other. An example is the prisoners' dilemma.
- *Coordination games*: There are two pure strategies and a mixed strategy that correspond to Nash equilibria. Only the pure strategies survive in an evolutionary setting. Generic examples are the traffic coordination problem and the stag hunt.
- *Anti-coordination games*: The only evolutionarily stable strategy is the mixed strategy. This category is exemplified by the hawk-dove game.

The hawk-dove game is the paradigm model of conflict over limited resources, and this is also the game that was chosen by Maynard Smith and Price in a seminal paper as an illustration of "the logic of animal conflict".[17] Doves are cooperative and share resources equally, resulting in equal payoffs of $G/2$, whereas hawks fight to gain full control. If a dove meets a hawk, it retreats, and the hawk reaps the full benefit (G). If a hawk meets a hawk, they share equally but have to pay the cost of conflict, C, assumed to be greater than G. This yields the game matrix in Fig. 6.2.

In repeated games, it is possible to enrich the spectrum of strategies by conditioning the choice on the other player's behaviour. The strategy *bully* starts by being aggressive and responds to cooperative behaviour by being hawkish, but it retreats if it meets with repeated resistance. Its mirror image is *retaliator*, which starts with cooperation but responds to aggressive behaviour.[18]

Fig. 6.2 The hawk-dove game

	Hawk	Dove
Hawk	(G-C)/2, (G-C)/2	G, 0
Dove	0, G	G/2, G/2

---

[16] Milanovic (2013), Sect. 4. The underlying data base covers 151 civil wars in 70 countries between 1945 and 2002 (Sambanis and Schulhofer-Wohl 2009).

[17] Maynard Smith and Price (1973).

[18] Ibid.

In the single-population game, the only evolutionarily stable outcome is the mixed strategy where every player chooses a mixture of hawk and dove, the weights given by the parameters G and C. In a two-population game, the situation is radically different, however. The reason is that any small perturbation from the stationary point corresponding to the mixed-strategy equilibrium will result in a specialisation of the two populations into hawks and doves.[19] This pattern is analogous to what was observed in the previous section on the difference between homogeneous and heterogeneous populations.

This stylised model may be developed and refined in a number of directions, and such amendments will in general modify the conclusions. In the basic form, all hawks and doves are assumed to be equal in capacity; if hawks vary in this respect and if the capacity of each hawk is known to the player itself, the evolutionarily stable level of fighting will be reduced.[20]

Many interesting applications involve *more than two players*. N-person variations give rise to new patterns of interaction. For instance, if doves are capable of organising some form of protection against exploitation at a cost, this increases the likelihood of non-aggressive behaviour.[21] A multiplicity of solutions are possible, ranging from hawk dominance via coexistence and bi-stable modes (where the outcome depends on initial conditions) to dove dominance.[22]

The basic game is completely abstract in its lack of *spatial dimensions*. Various developments have been made, such as playing the game on a surface or in a network. Spatial games were introduced by Nowak and May and turn out to display behavioural patterns that are sometimes both quantitatively and qualitatively different from the original games.[23] Cooperative behaviour in the prisoners' dilemma can benefit from the spatial structure if cooperative agents are able to form pockets of cooperation in an otherwise non-cooperative world.[24] This also holds for the hawk-dove game.[25] Playing the game on a surface yields a proportion of hawks that is generally lower than in the original game. Retaliator is also a more successful strategy in the spatial version.

The hawk-dove game has also been studied on *networks* with structures varying from regular lattices to random graphs.[26] In these structures, the outcome is ambiguous. Cooperation among doves can benefit or be hindered, depending on update rules and cost assumptions.

---

[19] Weibull (1995), pp. 183f.

[20] McNamara and Houston (2005).

[21] Chen et al. (2017).

[22] Similar qualitative changes appear in the transition from 2-person to N-person prisoners' dilemmas; see, for example, Molander (1992).

[23] Nowak and May (1992).

[24] Bowles (2008) has coined the term *conflict as the midwife of altruism*. For a broad overview of the collective-action problem among nonhuman primates, see Willems and van Schaik (2015).

[25] Killingback and Doebeli (1996).

[26] Tomassini et al. (2006).

Uncertainty is another factor that, if taken into account, may affect the outcome. Random variations in space and time that are of the same order of magnitude as the payoffs in general facilitate aggressive behaviour.[27] These disturbances represent an obstacle to cooperative agents that try to form closed areas of cooperation. This effect of uncertainty may be part of the explanation for aggressive behaviour both in animal and human populations.[28]

6.3 Conformity, Segregation and Imitation

Conformity in human behaviour can have different sources. One is a basic search for similarity. Human beings have a drive to socialise with people who are similar to themselves in some respect. This tendency towards *homophily* affects our choice of connections in a variety of networks, such as family and marriage, friendship, neighbourhoods, work and membership in various organisations.[29] Social networks in real life are therefore more homogeneous than would be the case if they formed randomly. This limits people's information base, their attitudes, and their actions and interactions.

A second source of conformity is the search for *improvement*. People observe their neighbours in a variety of contexts and sometimes mimic them if they expect improvements by doing so. This may refer to education, place of residence, fertility choices, clothing or any other aspect of life. This source of conformity is of obvious interest to the economic profession, whereas the homophilic tendency has been studied mainly by sociologists.

Both these categories of collective behaviour are relevant to issues of distribution and inequality. The segregating patterns generated by homophilic tendencies are not linked to inequality by necessity, but in practice they often turn out to be. The search for improvements also tends to generate inequality, because of differences in information bases, economic restrictions or other obstacles.

Herd Behaviour
In situations where information is sparse, a common rule of thumb is to follow in the footsteps of others. A person armed with scant information who faces a choice may assume that someone else who has already made a choice was better informed. This creates the problem of weighing one's own information against the assumed information underlying the other person's choice, which is difficult in a situation characterised by high uncertainty. The power of the concrete example is of course exploited in marketing, where well-known people—athletes, film stars, etcetera—

---

[27] Perc (2007).

[28] There are also parallels between the hawk-dove game and the prisoners' dilemma with respect to uncertainty. The introduction of uncertainty renders certain previously efficient strategies highly inefficient; see Molander (1985).

[29] See the survey by McPherson et al. (2001).

are used as models in order to convince potential customers of the superiority of a particular brand. But even in the absence of such persuasive tricks, the reflex of following the example is strong.

Various analyses of this decision problem have come to similar conclusions—that pure herd behaviour is dysfunctional and may lead a group to completely erroneous conclusions at the aggregate level.[30] This outcome is possible even under the assumption of (boundedly) rational behaviour, that is, when no particular value is assigned to conformity itself. Firstcomers will play a disproportionately important role, and the risk of volatile group behaviour is high. The main antidote to this behaviour is some ad hoc rule, such as deliberately choosing an alternative that may not appear to be optimal, either by randomisation or by actively going against the stream.[31]

The importance of this herding reflex to the problem of inequality derives from the fact that information is not uniformly distributed in a society, and that access to high-quality information that reduces the risk of mistakes can be bought. A person who has a well-informed peer group is also in a better position, even though he or she need not be objectively better informed.

Segregation

Ethnic origin is the strongest source of homophilic drives, followed by age, religion, education, occupation and gender.[32] The arenas where these drives are formed and express themselves are found in social networks and in geographical space. The result is more or less pronounced forms of segregation.

The classical reference for the modern analysis of segregation is due to Thomas Schelling.[33] A two-dimensional space is populated by two groups, whose members have certain preferences concerning their neighbourhood, such as not belonging to a small minority in this local neighbourhood. If they are dissatisfied, they move. Such moves can in turn release other moves, and the question to be answered is what possible stable equilibria of this sequential game look like.

A first important conclusion from the analysis is that even relatively weak requirements regarding own-group representation can generate highly segregated patterns of settlement. There is no simple correspondence between individual incentives and collective outcomes. An outside observer making inferences about individual preferences on the basis of aggregate patterns would most likely overstate the desire for "purity" in the local neighbourhood.

A second observation of general interest is the appearance of tipping points. Small changes in behavioural parameters can lead to large re-allocations, because of the appearance or disappearance of intersections between the curves that summarise the distribution of preferences in the two groups.

---

[30] See Banerjee (1992) and Arthur (1994), in particular Chaps. 5 and 8.

[31] Morone (2012).

[32] McPherson et al. (2001).

[33] Schelling (1971, 1978).

This sensitivity to small changes also concerns initial conditions, or prehistory. In this respect, there are parallels between this game of segregation and the games of conflict analysed in the previous section.

Schelling's model has been further developed and found to be robust under changing the assumptions on residents' preferences or the process of moving.[34] Even if the residents have a strict preference for integration, segregation may eventually prevail.

Participation and Imitation

Imitation or conformity—whatever makes the individual observe the choice of others in the population and adapt—implies that the standard assumption of independent, individual decision-making must be abandoned. In a situation where a binary choice has to be made—to contribute to a public good or not, to join or not to join an organisation—a simple decision rule is to contribute if a sufficient number of others do. Individuals differ in their requirements on "sufficiently", however, and the distribution function relative to this requirement will determine the range of possible outcomes.

Schelling established the prototype model for this type of conditional behaviour by observing that equilibria for the interactive process must be consistent, in the sense that requirements regarding the behaviour of others in the population must be compatible with their actual behaviour.[35] This is visualised in Fig. 6.3.

Possible equilibria are points of intersection between the cumulative function and the straight line, $y = x$. In cases A and C, there is only one equilibrium, corresponding to high and low levels of participation, respectively. For dynamic stability, it is also necessary that the curve intersects the straight line from above. In case B, the high- and low-level equilibria are stable, whereas the intermediate one is not.

Fig. 6.3 Schelling curves describing the cumulative function of requirements regarding other participants that must be satisfied for a subject to participate. Source: Schelling (1978)

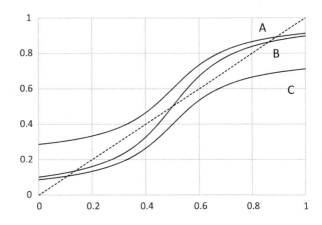

---

[34] Pancs and Vriend (2007).

[35] Schelling (1978).

As an alternative, it is possible to describe this behaviour in terms of utilities. Consider a binary choice of the above type, and denote by ω_i the choice of agent i, where $\omega_i = 0$ represents no action and $\omega_i = 1$ participation. Agent i attaches a proper value h_i to participating but in addition a value J_{ij}, if agent j also participates.

In the case of imitation, the goal of the players is similar in structure, and here the collective term can be set proportional to $(\omega_i - \omega_j)^2$, expressing the discomfort of deviating from others.

Let $\vec{\omega}$ denote the vector of choices made by the agents. The total utility for agent i can then be written

$$u\left(\omega_i \mid \vec{\omega}\right) = h_i\omega_i + \left\{\sum_{j\neq i} J_{ij}\omega_i\omega_j\right\} \tag{6.1}$$

Strictly speaking, the vector $\vec{\omega}$ is not known at the time of the decision, so the expression refers to expected rather than actual values. Assuming that the process has converged, these will also be assumed to coincide.

Different cases are possible. If h_i is positive, agent i will go ahead irrespective of the choices made by others. This corresponds to the left part of curve A in the figure. Assuming that the coefficients J_{ij} are positive, others will successively join as the number of participants increases. At the right end of the distribution, there may be agents for which the combination of proper value and collective value never becomes positive, and who will therefore never join.

Looking first at the two-person case, an interesting situation occurs when the proper value h_i is negative for both parties but $J_{ij} = J$ is positive. If $|h_i| > |J_{ij}|$, the game is a prisoners' dilemma, and the prospects for cooperative behaviour are at best uncertain. If the reverse inequality holds, the players face a stag hunt, and conditions are more propitious to cooperation.

For a dynamic description of the system, one can assume that each agent compares his or her current payoff with what others get, or to the average. Assuming that the rate of change is determined in relation to the average payoff leads to the replicator equation.

For larger populations, it is reasonable to introduce some uncertainty into the equation. The choice of agent i in this case is not given directly by maximisation of the utility function. Instead, the player chooses to participate with a certain probability, which can for instance be given by the logit distribution:

$$Prob(\omega_i = 1) = 1/\left[1 + \exp\left(-\beta\left(\vec{\omega}\right)\right)\right]$$

Here, the function $\beta(\cdot)$ is proportional to the utility function; if the constant of proportionality becomes very large, this is equivalent to the deterministic choice of maximising the utility function.

The general solution of the resulting system of equations is complicated. In special cases, the mathematics become more tractable. Assume for instance that interaction is symmetric and identical for all participants ($J_{ij} = J/(N-1)$), where N is the number of players), and further that they only want to conform with the average value ($=m$).[36] Then it can be shown that there exist one or three solutions that guarantee consistency, the number depending on the parameters of the problem.[37]

This result refers to global interaction. If instead it is assumed that each agent interacts only with the neighbours, local solutions that differ may emerge.[38] This takes us back to the problem of segregation and opens up rich possibilities for spatial variation—local homogeneity combined with diversity in the aggregate.

What connects this picture with the complex of inequality is precisely this diversity. In the typical case, the different equilibria that emerge and stabilise in the subpopulations represent different levels of well-being, corresponding to differences in levels of cooperation, trust, unemployment, crime rates, and so on. Members of a subpopulation may individually realise the dysfunctional character of their equilibrium but are—in the absence of successful collective action or assistance from outside—unable to change the current state of affairs.

6.4   Networks

Network structures, by their very nature, lend themselves to analysis using mathematical models. On the other hand, data collection in social networks, which are in focus here, is in most cases cumbersome. In technical systems, the structure and the character of exchange is reasonably precise, so some popular paths followed by researchers in the field have been either to focus on the infrastructure as such—for instance, the World Wide Web—or to use such a structure as a proxy for social networks, for example via e-mail statistics. A third alternative is to use the network concept in a more metaphorical sense.

Degree Distribution
From the perspective of inequality, the number of links connected to a node—the *degree* of the node—is a relevant entity. Having many links to other nodes is intuitively an asset and an approximate measure of power.

Consider a network with its structure of nodes—agents of various kinds—and links. Common questions asked concern how the number of links connected to a node is distributed, whether nodes are clustered or evenly connected, whether degrees are correlated and how resilient the network is to the disappearance of links. A standard approach is to imagine how the network was born and how it has grown and grows.

---

[36] This is referred to as *the mean-field approximation*.

[37] Brock and Durlauf (2001).

[38] See Blume and Durlauf (2001).

Assume first a *randomly* formed network with n nodes and the probability of a link $= p$. The distribution of degrees p_k will then be binomial:

$$p_k = \binom{n-1}{k} p^k (1-p)^{(n-1-k)}$$

As n becomes large, the binomial distribution can be approximated by the Poisson distribution. Setting $z = p(n-1)$, the asymptotical form of the distribution will be

$$p_k = z^k e^{-z}/z!$$

The Poisson distribution is skewed to the right, although relatively mildly so. The degree of skewness decreases with z. As p grows, the network becomes increasingly connected. The diameter d of a graph is the maximum distance between any two nodes. For random networks, d can be shown to be close to $ln(n)/ln(pn)$. The following typical results hold[39]:

- If $p{\cdot}n < 1$, the network will be composed of a number of isolated subnetworks.
- If $p{\cdot}n > 1$, a giant cluster, comprising most of the nodes, will develop.
- If $p > ln(n)$, the network is totally connected.

A radically different picture emerges, if links are formed not randomly but on the basis of the existing degree distribution. Assume that the network structure and the distribution are given at an instant in time t and further that the probability that a new node connects to an existing one is higher if the latter has a high degree, so-called *preferential attachment*. A natural assumption is that the probability of choosing a node i is proportional to its current degree k_i:

$$\mathcal{P}(choosing\ i) = k_i / \sum_j k_j$$

The asymptotic behaviour following this assumption can be derived in different ways. Assume, for instance, that nodes are added at a constant rate c per time unit and that k_i is added at time t_i. If k_i is approximated by a continuous variable, the following differential equation results:

$$\frac{dk_i}{dt} = c \cdot k_i / \sum_j k_j$$

---

[39] Albert and Barabási (2002), § III.E.

The sum in the denominator is taken over all nodes except the new one and will therefore equal approximately *2ct* (each node will be counted twice). The resulting differential equation will consequently take the simple form

$$\frac{dk_i}{dt} = k_i/2t$$

The solution is

$$k_i(t) = c\sqrt{t/t_i}.$$

In order to derive the distribution function for $k_i(t)$, note that

$$P(k_i(t) < k) = P\left(t_i > t\left(\frac{c}{k}\right)^2\right)$$

But the distribution $p(t_i)$ is uniform = *1/t*, whence the frequency function of k for large k satisfies

$$p(k) \sim k^{-3}$$

This type of power dependence implies that the degree distribution is *scale-free*.

In conclusion, the assumption of steady growth in combination with preferential attachment yields a distribution with a fat tail that is different both from the Poisson distribution of random networks and from the lognormal distribution, which emerges in the case of proportional growth according to Gibrat's rule. Both the growth assumption and preferential attachment of some sort are necessary for this conclusion.[40]

Other functional forms than proportionality between the probability of attachment and current degree are possible, sublinear or superlinear. In the latter case, the resulting distribution will be even more skewed to the right.

Empirically, distributions tend to fall somewhere in between these limits, defined by random growth and preferential attachment. The World Wide Web and the Internet fit the scale-free alternative well, whereas scientific collaboration and citations appear to be hybrids between scale-free and exponential or lognormal distributions.[41]

Centrality and Power

Influence or power is not determined solely by degree. A network has a topological structure, and it is the combination of degree and position in the network that decides whether an agent is powerful—a *key player*. Sociologists have used concepts such as

---

[40] Albert and Barabási (2002), § VII.C.
[41] Newman (2003), Fig. 6.

prestige or *centrality* in order to describe this basis of power.[42] These measures have been further developed by Ballester et al. using a model similar to the one in the previous section when analysing participation (Eq. 6.1).[43]

Consider a number of agents, denoted by i, who choose a variable x_i. Agent i has a utility function u_i, defined as follows:

$$u_i\left(x_i, x_{j\neq i}\right) = a_i x_i - b_i x_i^2 / 2 + \sum_{j\neq i} c_{ij} x_i x_j$$

Here, a_i and b_i are positive, whereas c_{ij} can have different signs, depending on whether interaction is positive or negative. A condition for optimum is that the derivative be zero, or

$$x_i = \frac{a_i}{b_i} + \sum_{j\neq i} \frac{c_{ij}}{b_i} x_j$$

This defines a system of equations for the x_i's. Define $\vec{\alpha}$ as the vector of a_i/b_i and let $g_{ij} = c_{ij}/b_i$ be the elements of the matrix G. If the b_i's are sufficiently large, the matrix $(I - G)$ is invertible, and the solution is given by

$$\vec{x} = (I - G)^{-1}\vec{\alpha} = \sum_{i=0}^{\infty} G^i \vec{\alpha}$$

Le d_{ii} be the diagonal elements of $(I - G)^{-1}$. Ballester et al. define an *intercentrality measure* m_i in the following way:

$$m_i = b_i^2 / d_{ii}$$

The main result is that the key player of the network—the one whose removal affects the aggregate activity the most—is the one with the highest intercentrality measure. This generalises Bonacich's centrality measure,[44] which only considers the direct effect of a player. The intercentrality measure also includes the indirect effects of a player via the network. In general, these two measures give different answers to the question who is the most important or powerful player.

There may in principle be more than one key player, and the key player need not be the most active one. Removing the key player may be significantly more efficient than removing the most active player in cases of adverse behaviour.[45]

---

[42] Prestige: Katz (1953), centrality: Bonacich (1987). For a survey of the key-player literature, see Zenou (2016).

[43] Ballester et al. (2006).

[44] Bonacich (1987).

[45] See Boguslaw (2017), Chap. 3, for a real-life example (disruptive behaviour in class-rooms).

6.5   Markets

In Sects. 5.1 and 6.1, the distribution of the aggregate surplus between capital-owners and labour was analysed using a two-party interaction description. We now consider the problem of distribution from the perspective of a large number of individuals interacting via markets for capital and labour and ask the question about the distribution of incomes—which may come from both labour and capital for one and the same person—and how it can be derived. Income distributions exhibit thick tails, that is, with a relatively high fraction of high-income earners, and a model for income and wealth formation should be able to reproduce this stable observation.[46]

Random Walks
Champernowne showed in an early paper that the Pareto distribution can be reproduced if income movements are modelled as a random walk between different income strata.[47] This model, further developed by Mandelbrot,[48] builds on the assumption that incomes are subject to a downward drift but that there is a reflecting barrier at the lower end of the income ladder. Both Champernowne and Mandelbrot abstained from giving an economic rationale for their model, which is rather to be considered as a description of a mechanism that generates the desired result. As remarked by Stiglitz, Champernowne's conditions can be shown to be satisfied in a balanced-growth model, if the rate of return on capital is random and the capital-to-labour ratio is sufficiently high.[49]

Balanced-Growth Distributions
In the study by Stiglitz just cited, the purpose is to identify economic forces that tend to equalise wealth and forces that make it less evenly distributed. Some different assumptions are examined concerning, for instance, the form of the consumption function, the heterogeneity of labour, and inheritance policies.

In the base case, it is assumed that all workers receive the same wage, and that each factor (labour and capital) is paid according to its marginal product (wage rate w and interest rate r, respectively). In this way, differences in income stem from differences in capital assets. The aggregate production function f is concave and exhibits constant returns to scale. If y is output per capita and k is the aggregate capital-to-labour ratio, then

$$y = f(k); f'(k) > 0; f''(k) < 0$$

We have

---

[46]For a survey of macroeconomic models of distribution and growth, se Bertola (2000).

[47]Champernowne (1953).

[48]Mandelbrot (1961).

[49]Stiglitz (1969).

$$r = f'(k); w = f(k) - kf'(k)$$

Assume now that the population consists of groups denoted by i and let c_i be the capital asset per capita:

$$y_i = w + rc_i$$

Different assumptions on the savings rate are possible. In the base case, a linear relationship between income and saving is assumed:

$$s_i = b + my_i$$

Reproduction rates n are assumed to be constant across groups, implying that the fraction of each group a_i in the population remains constant.

Under these assumptions, the aggregate rate of capital accumulation in society is independent of the distribution of capital:

$$\dot{k} = \sum_i a_i \dot{c}_i = b + m(w + rk) - nk$$

In a balanced growth regime, $\dot{k} = 0$, so $my = nk - b$. Because of the concavity of the production function, if $b > 0$, there is only one solution, corresponding to a globally stable growth path. If $b < 0$, there are two solutions, the left one (k^*) unstable and the right one (k^{**}) stable. In consequence, a motion starting to the left of k^* implies a continuously falling k, whereas a motion starting above this value will converge to the upper equilibrium.

In order to examine the distributional consequences of these two scenarios, it suffices to consider the case of two classes, 1 and 2. For the relative growth rates of the capital assets of these two groups, the following equation holds:

$$\frac{\dot{c}_1}{c_1} - \frac{\dot{c}_2}{c_2} = (n - mr)k\left(\frac{1}{c_1} - \frac{1}{c_2}\right)$$

In the *upper* aggregate equilibrium (k^{**}), $(n - mr) > 0$, so $c_1 < c_2$ implies that c_1 grows faster than c_2 and that capital assets will be equalised in the long run. In the *lower* equilibrium (k^*), by contrast, the capital-to-labour ratio is unstable both in the aggregate and with respect to relative assets. If the assets of the two classes are perturbed in such a way that the overall k is conserved, the assets of the class with lower assets are reduced, whereas the reverse holds for the upper class, and the difference between the classes will grow over time.

It can also be shown that there is a critical value $\hat{k}$ between k^* and k^{**} such that an economy starting between $\hat{k}$ and k^* will first experience an increase in inequality and eventually a decrease and convergence to equal assets.

The realism of the model can be increased in various ways. Two assumptions appear particularly problematic: the homogeneity of the labour force and linear savings behaviour. If heterogeneity is permitted but linearity is maintained, then the distribution of capital assets in equilibrium will be determined by the distribution of productivity: lower productivity implies lower assets, as expected. For sufficiently low productivity levels, the less productive will be indebted. This case consequently represents a growth path characterised by increasing inequality.

Relaxing the linearity of savings is even more important. As we have seen, wealthy households save relatively more,[50] so a convex savings function is a step towards increased realism. In the convex case, there can be two groups that are in equilibrium, satisfying the stability condition $s(y) = nc$. Groups with lower incomes will save less and have smaller assets, and inequality will rise over time.

In summary, this highly stylised model, under reasonably realistic assumptions, will predict that inequality will rise over time, in conformity with observations of economies working under normal conditions.

The Importance of Risk Aversion

In the above model, heterogeneity in productivity may be a source of rising inequality. Another form of heterogeneity refers to *patience*, as expressed by the individual discount factor. Becker has shown that in the long run, the income distribution in a society with heterogeneity in this respect will be determined by the lowest discount factor.[51] The household with the lowest discount factor will in the long run own all the capital, whereas other household incomes consist of wages only. Given that wealthy households can afford to wait longer, there is a possibility that such a development arises naturally, as a result of random income fluctuations. This type of analysis has been carried out by Fernholz and Fernholz, not with reference to patience but to risk aversion.[52]

The hypothetical society studied consists of households that are equal in all important respects—productivity, risk aversion, discount factor, etc. They all work and receive the same wage λ, and they invest their surplus in a risk-free asset with payoff r or a risky asset, whose payoff follows a Brownian motion in continuous time:

$$dP_i(t) = \alpha P_i(t) + \sigma P_i(t)dB_i(t)$$

Here, $\alpha > r$ is the average payoff, and σ the standard deviation of instantaneous payoff.[53] The risk is idiosyncratic, and markets are assumed to be incomplete, so that this risk cannot be insured. At time t, each household consumes $c_i(t)$ and invests a

---

[50] Dynan et al. (2004), Battistin et al. (2009).

[51] Becker (1980).

[52] Fernholz and Fernholz (2014).

[53] Actually, because of the asymmetry of the Itô integral, the equation presented does not model a process with growth rate α but $(\alpha-\sigma^2)/2$. In order to symmetrise the definition, a correction term is necessary, the so-called Stratonovich correction (Stratonovich 1966). The effect under study here derives from the noise term and not from the growth term, however, so conclusions are unaltered by this adjustment.

fraction $\varphi_i(t)$ of its current wealth $w_i(t)$. In terms of risk aversion, constant relative risk aversion is assumed; there is empirical evidence for this assumption.[54]

The optimisation problem faced by each household is consequently

$$J(w,t) = \max_{c_i(t),\,\varphi_i(t)} E\left[\int_t^\infty c_i(s)^{(1-\gamma)} \cdot e^{-\rho s}/(1-\gamma)ds\right], \gamma < 1$$

Here, γ is the constant of relative risk aversion, and ρ is the discount factor. This optimisation problem can be solved explicitly:

$$x_i(t) = x_i(0)\, exp\left[(\Lambda - \tfrac{1}{2}\Gamma^2)t + \Gamma B_i(t)\right],$$

where Λ and Γ are constants determined by the parameters of the problem. This implies that the wealth of each household evolves as a geometric Brownian motion with the same expected growth rate. There is no equalising mechanism in this economy.

Now let $\theta_{max}(t)$ be the share of the wealthiest household of total wealth at time t. Fernholz and Fernholz then prove that

$$\lim_{T\to\infty}\frac{1}{T}\int_0^T \theta_{max}(s)ds = 1 \; almost \; surely.$$

In plain words, this implies that in the long run, the time-averaged share of the wealthiest household converges to 1, that is, the total wealth in the economy will be concentrated to one household. This is true in spite of the fact that all individuals are assumed to have identical abilities, identical patience and the same expected returns on their risky investments. The playing field is level: they start with the same assets, and there is equal opportunity between the households. It is worth noting that it is not necessarily the same household that is at the top of the wealth ranking all the time; because of the stochastic element, recurrent changes of ranking are possible.

In summary, this idealised economy exhibits *symmetry breaking*. A large group of identical individuals, interacting under common and non-discriminating rules, converge towards a state that is highly asymmetrical.

This is a stark result, but the effect modelled is relatively weak viewed in isolation; the transition from the initial state of equality to one characterised by high inequality is a matter of centuries with realistic assumptions on the parameters involved. In real life, other effects would work in the same direction, however, such as qualified advice on asset management or insider information supplied to wealthy households. The dynamic behaviour would be similar, and the time constant significantly smaller.

---

[54] Meyer and Meyer (2006) observe in their survey that there is strong support for decreasing *absolute* risk aversion and against decreasing *relative* risk aversion. The limiting case, constant relative risk aversion, is supported by a more recent analysis (Chiappori and Paiella 2011, Italian data).

Wealth Versus Capital

The article by Stiglitz cited above was written during an era when industrial economies were growing at a steady rate, inequality was decreasing, and both the capital-to-labour ratios and their relative share of the value added produced were stable. Focus was on explaining this stable pattern, and the Kaldor-type model with variations was the main tool. Half a century later, the stylised facts to be explained turned out to be radically different, for instance:

- Inequality in both wages and wealth was increasing.
- Average wages had stagnated, in spite of increasing productivity.
- The ratio between wealth and income was increasing.
- Return on capital was stable, in spite of the increasing wealth-to-income ratio.

A natural question to ask is whether the same basic analysis tools are usable in the new situation. As we have seen, there is a tendency in some set-ups for the Kaldorian framework to gravitate towards stable and egalitarian solutions, but non-egalitarian equilibria emerge when assumptions about labour productivity and saving behaviour are more realistic than in the base case. Building on the previous framework, Stiglitz has developed the model in a number of directions,[55] notably:

- The crucial step is to make a distinction between wealth and capital. Not all wealth is a productive factor in the economy. There is consequently a need for introducing a second type of asset, provisionally called *land*.
- Savings in the majority of the population are mainly for old age, but there is a small category of capital-owners who aim to create financial holdings for the transmission across generations.
- The economy is not fully competitive.
- Land (in the above, generalised sense) can be used as a collateral, which on the one hand contributes to inequality in the financial markets, but on the other can also be affected by public policies.

An indication of a gap between wealth and capital follows directly from national account statistics on growth and savings rate. Using the common notation Y for total output, K for aggregate capital, s for net saving rate and g for growth rate, we have

$$\frac{d}{dt} \log (K) = \frac{sY}{K}; \frac{d}{dt} \log \left(\frac{K}{Y}\right) = \frac{sY}{K} - g$$

In the United States during the period 1970–2010, the average net savings rate amounted to 5.2 per cent, and the average growth rate was 2.8 per cent. The wealth-to-income ratio has varied between 4 and 4.6, so according to the second equation, the wealth-to-income ratio should have declined by 1.5–1.6 per cent, if capital (K) had been equal to wealth (W). There is a *wealth residual*, somewhat reminiscent of

---

[55] Stiglitz (2016).

the growth residual discovered by Solow in the 1950s, when only a minor fraction of total growth could be explained by increased capital input.

A number of different sources can be found for this residual:

- rents on land and other assets that are not being produced, such as housing in attractive locations, art objects and positional goods
- control of resources by increasing market power,[56] various forms of exploitation in the financial as well as other consumer sectors (insider trading, predatory lending, robot trading, exploitation of systematic irrationalities in risk taking, etc.)
- corporate rent seeking, as illustrated for instance by government subsidies to the financial sector and the nuclear power industry[57]
- other rents on knowledge and information, such as interbank trading and very large data-bases storing personal information.

If the classical production is expanded with a term representing land (in the broad sense), one gets

$$Y = F(K, L) + R,$$

where Y is total output, K and L capital and labour, and R is the return on land, assumed fixed. Let F_K be the return to capital and μ the depreciation rate, and let p be the price of the (unproductive) land asset. In equilibrium, the return to both types of assets must be the same:

$$\frac{d}{dt} \log (p) = (F_K - \mu) \tag{6.2}$$

For simplicity, it is assumed that only large capital-owners save and that they save a constant fraction s of their total income. This yields a second equation:

$$\frac{dK}{dt} = (sK - (1 - s)p)(F_K - \mu) \tag{6.3}$$

The right-hand sides of (6.2), (6.3) are zero for

$$F_K = \mu; \quad p = sK/(1 - s)$$

Let (K^*, p^*) be the solution of this system of equations. Then the solution of the system of differential Eqs. (6.2), (6.3) can be summarised as follows:

- Possible stationary points are the segment of the line $K = K^*$ satisfying $p < p^*$.
- All motions starting from points $K > K^*$ and a subset of the area $(K < K^*, p < sK/(1-s))$ will converge to a stationary point on this segment.

---

[56] See De Loecker et al. (2020) and Syverson (2019) for some views on this topic.
[57] Kelly et al. (2016) and Dubin and Rothwell (1990), respectively.

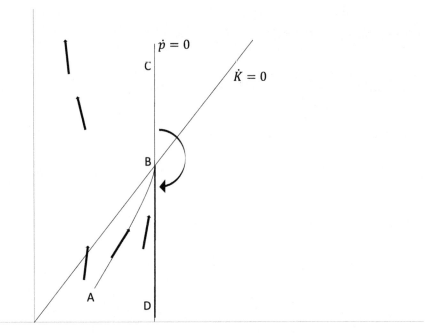

Fig. 6.4 Steady states and dynamics with non-productive capital (land). Source: Stiglitz (2016)

- Motions starting in the remaining part of the first octant will not converge. K will tend to 0 and p to ∞. This case is equivalent to a speculative bubble in the non-productive assets.

The motion is sketched in Fig. 6.4.

If a more realistic, nonlinear savings function is assumed, more complex behavioural patterns are possible. The motion may be unstable with K or p tending to ∞, or there may be a limit cycle, representing a quasiperiodic motion in K and p. Other assumptions concerning the non-productive asset are possible.

In summary, the distinction between wealth and capital and the analysis sketched above explains part of the increasing inequalities in industrial societies—the increasing gap between measured wealth and productive capital, stagnating wages in spite of increases in productivity and GDP, an increasing focus on positional goods and the emergence of speculative bubbles. The market economy is inherently unstable.

6.6   Firm Size Distribution

The distribution of firm sizes has been widely studied in economics, both with respect to its determinants and to its consequences. This distribution is not of direct relevance to inequality considerations—efficiency rather than equity would be the main concern in most policy discussions on this topic—but there are indirect effects on wages and profits that merit attention.

Firm size was one of the examples that Gibrat discussed in his classical paper,[58] and he deduced a lognormal distribution based on the assumption of proportional growth. Empirical studies of firm sizes differ somewhat in their conclusions. Some report a lognormal distribution,[59] while others reproduce a Pareto distribution,[60] or even a Pareto distribution with a decreasing exponent, indicating a thickening tail and an increasing concentration.[61] There are both differences and similarities between sectors.[62]

Various explanations for the thick tail have been advanced. Most of them rationalise this behaviour by referring to economies of scale, for instance with respect to research and development,[63] but there may also be a direct effect of size, resulting in market power.[64]

The link to income distribution goes via this market power, whatever its origin.[65] A study of a choice of OECD countries indicates substantial, regressive distributional effects of market power.[66] The wealth of the uppermost decile group is increased by 12 to 21 per cent for a reasonable assumption on saving behaviour, and the income of the lowest quintile group is reduced by 14 to 19 per cent. Although caveats are necessary for these sorts of estimates, even lower-bound assumptions yield considerable effects.

6.7 Summary

The family of models that have been presented in this chapter is variegated, but an attempt to summarise some general observations is nonetheless warranted.

- The conclusions drawn from model-based analyses are sometimes robust, sometimes sensitive to assumptions made. For instance, assumptions on homogeneous labour and linear savings in a standard, balanced-growth model tend to yield convergence of assets, while a more realistic assumption leads to increasing inequality. It takes a specific study in each case to judge on robustness.

---

[58] Gibrat (1930, 1931).

[59] Cabral and Mata (2003) (Portugal).

[60] Axtell et al. (2001) (US), Coad (2010). The transition from a lognormal to a Pareto distribution depends on the assumptions made, and the estimation may be problematic (Clauset et al. 2009).

[61] Gao et al. (2015) (China).

[62] Bottazzi and Secchi (2003).

[63] Pagano and Schivardi (2003).

[64] Ghemawat (1990).

[65] Autor et al. (2017).

[66] Ennis et al. (2017).

- Bilateral negotiations are sensitive to the fallback positions of the parties involved. They exhibit instability in the case of repeated meetings and spontaneously evolve into asymmetric relationships. More pronounced conflict situations similarly tend to develop into asymmetric equilibria. Aggression may be an important factor affecting the outcome of such exchanges.
- A spatial dimension makes local solutions possible, which creates diversity at the aggregate level. Whether this diversity is welfare enhancing or not depends on the circumstances.
- Uncertainty has an ambivalent effect on outcomes. In some situations, it may increase the likelihood of cooperative behaviour and improve welfare; in others it may lead to increased aggression and less cooperation.
- Many decision problems in groups are solved by imitative behavioural strategies. Such imitation or herding behaviour often opens up for multiple equilibria, sensitivity to initial conditions and abrupt changes resulting from small variations in the environment. Different equilibria correspond to different levels of welfare, which represents yet another source of inequality.
- Networks tend to generate inequality, more or less pronounced, depending on mechanisms for link formation and growth. The power attached to a node depends both on the number of links and the position in the network. Identifying key players with more-than-average power requires detailed analysis.
- Market models are somewhat ambivalent, but realistic assumptions tend to predict that inequality will increase with time. Such tendencies may be triggered by differences among individual agents (productivity, patience, etc.), but there are cases where inequality will develop even when starting from fully symmetrical initial conditions (symmetry breaking). The distinction between capital and wealth is essential. Thick-tailed distributions of firm sizes are yet another potential source of inequality among individuals via market power effects.

Bibliography

Albert, R., & Barabási, A.-L. (2002). Statistical mechanics of complex networks. *Review of Modern Physics, 74*, 47–97.

Arrow, K. J. (1965). *Aspects in the theory of risk bearing*. Yrjö Johansson Saatio.

Arthur, W. B. (1994). *Increasing returns and path dependence in the economy*. The University of Michigan Press.

Autor, D., et al. (2017). Concentrating on the fall of the labor share. *American Economic Review: Papers & Proceedings, 107*(5), 180–185.

Axtell, R. L., et al. (2001). The emergence of classes in a multi-agent bargaining model. In S. N. Durlauf, & H. P. Young (Eds.), *Social dynamics*. Washington, DC: Brookings Institution Press/ Cambridge, MA: MIT Press.

Ballester, C., et al. (2006). Who's who in networks. Wanted: The key player. *Econometrica, 74*(5), 1403–1417.

Banerjee, A. V. (1992). A simple model of herd behavior. *The Quarterly Journal of Economics, 107*(3), 797–817.

Battistin, E., et al. (2009). Why is consumption more lognormal than income? Gibrat's law revisited. *Journal of Political Economy, 117*(6), 1140–1154.

Becker, R. A. (1980). On the long-run steady state in a simple dynamic model of equilibrium with heterogeneous households. *Quarterly Journal of Economics, 95*(2), 375–382.

Bertola, G. (2000). Macroeconomics of distribution and growth. Ch. 9 in A.B. Atkinson & F. Bourguignon (Eds.), *Handbook of income distribution*, Vol. 1. Amsterdam: North-Holland.

Blume, L. E., & Durlauf, S. N. (2001). The interaction-based approach to socioeconomic behavior. In S. N. Durlauf, & H. P. Young (Eds.), *Social dynamics*. Washington, DC: Brookings Institution Press/Cambridge, MA: MIT Press.

Boguslaw, J. (2017). When the kids are not alright. *Essays on childhood disadvantage and its consequences*. Ph.D. diss., Dept. of Economics, University of Stockholm.

Bonacich, P. (1987). Power and centrality: A family of measures. *American Journal of Sociology, 92*(5), 1170–1182.

Bottazzi, G., & Secchi, A. (2003). Common properties and sectoral specificities in the dynamics of U.S. manufacturing companies. *Review of Industrial Organization, 23*, 217–232.

Bowles, S. (2008). Conflict: Altruism's midwife. *Nature, 456* (20 November 2008), 326–327.

Bredemeier, H. C. (1978). Exchange theory. In T. Bottomore & R. Nisbet (Eds.), *A history of sociological analysis*. Basic Books.

Brock, W. A., & Durlauf, S. N. (2001). Discrete choice with social interactions. *The Review of Economic Studies, 68*(2), 235–260.

Cabral, L. M. B., & Mata, J. (2003). On the evolution of firm size distribution: Facts and theory. *American Economic Review, 93*(4), 1075–1090.

Champernowne, D. G. (1953). A model of income distribution. *The Economic Journal, 63*(250), 318–351.

Chen, W., et al. (2017). Evolutionary dynamics of N-person hawk-dove games. *Nature Scientific Reports, 7*, 4800. https://doi.org/10.1038/s41598-017-04284-6

Chiappori, P.-A., & Paiella, M. (2011). Relative risk aversion is constant: Evidence from panel data. *Journal of the European Economic Association, 9*(6), 1021–1052.

Clauset, A., et al. (2009). Power-law distributions in empirical data. *SIAM Review, 51*(4), 661–703.

Coad, A. (2010). The exponential age distribution and the Pareto firm size distribution. *Journal of Industrial Competition and Trade, 10*, 389–395.

Collier, P., & Hoeffler, A. (2004). Greed and grievance in civil war. *Oxford Economic Papers, 56*(4), 563–595.

De Loecker, J., et al. (2020). The rise of market power and the macroeconomic implications. *The Quarterly Journal of Economics, 135*(2), 561–644.

Dubin, J. A., & Rothwell, G. S. (1990). Subsidy to nuclear power through Price-Anderson liability limit. *Contemporary Policy Issues, 8*(3), 73–79.

Dynan, K. E., et al. (2004). Do the rich save more? *Journal of Political Economy, 112*(2), 397–444.

Ennis, S., et al. (2017). *Inequality: A hidden cost of market power*. www.oecd.org/daf/competition/inequality-a-hidden-cost-of-market-power.htm. Paris: OECD.

Fernholz, R., & Fernholz, R. (2014). Instability and concentration in the distribution of wealth. *Journal of Economic Dynamics & Control, 44*, 251–269.

Foster, D., & Young, H. P. (1990). Stochastic evolutionary game dynamics. *Theoretical Population Biology, 38*(2), 219–232.

Gao, B.-J., et al. (2015). On the increasing inequality in size distribution of China's listed companies. *China Economic Review, 36*, 25–41.

Ghemawat, P. (1990). The snowball effect. *International Journal of Industrial Organization, 8*, 335–351.

Gibrat, R. (1930). Une loi des répartitions économiques: l'effet proportionnel. *Bulletin de la Statistique Générale de la France, 19*, 469–516.

Gibrat, R. (1931). *Les Inégalités Économiques*. Librairie du Recueil Sirey.

Katz, L. (1953). Anew status index derived from sociometric analysis. *Psychometrika, 18*, 39–43.

Kelly, B., et al. (2016). Too-systemic-to-fail: What option markets imply about sector-wide government guarantees. *American Economic Review, 106*(6), 1278–1319.

Killingback, T., & Doebeli, M. (1996). Spatial evolutionary game theory: Hawks and Doves revisited. *Proceedings of the Royal Society of London B, 263*, 1135–1144.

Mandelbrot, B. (1961). Stable Paretian random functions and the multiplicative variation of income. *Econometrica, 29*(4), 517–543.

Maynard Smith, J., & Price, G. R. (1973). The logic of animal conflict. *Nature, 246*, 15–18.

McNamara, J. M., & Houston, A. I. (2005). If animals know their own fighting ability, the evolutionarily stable level of fighting is reduced. *Journal of Theoretical Biology, 232*(1), 1–6.

McPherson, M., et al. (2001). Birds of a feather: Homophily in social networks. *Annual Review of Sociology, 27*, 415–444.

Meyer, D. J., & Meyer, J. (2006). Measuring risk aversion. *Foundations and trends in microeconomics* 2(2), 107–203. Reprinted by Hanover, MA: now Publishers.

Milanovic, B. (2013). The inequality possibility frontier. Extensions and new applications. *Policy research working paper* 6449. Washington, DC: The World Bank.

Milanovic, B. (2018). Towards an explanation of inequality in premodern societies: The role of colonies, urbanization, and high population density. *Economic History Review, 71*(4), 1029–1047.

Milanovic, B., et al. (2011). Pre-industrial inequality. *The Economic Journal, 121*(551), 255–272.

Molander, P. (1985). The optimal level of generosity in a selfish, uncertain environment. *Journal of Conflict Resolution, 29*(4), 611–618.

Molander, P. (1992). The prevalence of free-riding. *Journal of Conflict Resolution, 36*(4), 756–771.

Morone, A. (2012). A simple model of herd behavior, a comment. *Economics Letters, 114*(2), 208–211.

Newman, M. E. J. (2003). The structure and function of complex networks. *SIAM Review, 45*(2), 167–256.

Nowak, M. A., & May, R. M. (1992). Evolutionary games and spatial chaos. *Nature, 359*, 826–829.

Pagano, P., & Schivardi, F. (2003). Firm size distribution and growth. *Scandinavian Journal of Economics, 105*(2), 255–274.

Pancs, R., & Vriend, N. J. (2007). Schelling's spatial proximity model of segregation revisited. *Journal of Public Economics, 91*(1–2), 1–24.

Perc, M. (2007). Uncertainties facilitate aggressive behavior in a spatial Hawk–Dove game. *International Journal of Bifurcation and Chaos, 17*(11), 4223–4227.

Pratt, J. W. (1964). Risk aversion in the small and in the large. *Econometrica, 32*(1–2), 83–98.

Sambanis, N., & Schulhofer-Wohl, J. (2009). What's in a line? Is partition the solution to civil war? *International Security, 34*(2), 82–118.

Schelling, T. (1971). Dynamic models of segregation. *Journal of Mathematical Sociology, 1*, 143–186.

Schelling, T. (1978). *Micromotives and Macrobehavior*. Norton.

Smith, A. (1776). *An inquiry into the nature and causes of the wealth of nations*. Several later editions.

Stiglitz, J. E. (1969). Distribution of income and wealth among individuals. *Econometrica, 37*(3), 382–397.

Stiglitz, J. E. (2016). New theoretical perspectives on the distribution of income and wealth among individuals. Ch. 1 in K. Basu, & J. E. Stiglitz (Eds.), *Inequality and growth: Patterns and policy. Volume I: Concepts and analysis*. Berlin: Springer.

Stratonovich, R. L. (1966). A new representation for stochastic integrals and equations. *SIAM Journal of Control, 4*, 362–371.

Syverson, C. (2019). Macroeconomics and market power: Context, implications, and open questions. *Journal of Economic Perspectives, 33*(3), 23–43.

Tomassini, M., et al. (2006). Hawks and Doves on small-world networks. *Physical Review* E 73, 016132 – Published 26 January 2006.

Vieille, N., & Weibull, J. W. (2008). Multiple solutions under quasi-exponential discounting. *Economic Theory, 39*, 513. https://doi.org/10.1007/s00199-008-0368-2

Weibull, J. W. (1995). *Evolutionary game theory*. MIT Press.

Willems, E. P., & van Schaik, C. P. (2015). Collective action and the intensity of between-group competition in nonhuman primates. *Behavioral Ecology, 26*(2), 625–631.

Young, H. P. (1998). *Individual strategy and social structure. An evolutionary theory of institutions*. Princeton University Press.

Zenou, Y. (2016). Key players. Ch. 11 in Y. Bramoullé et al. (Eds.), *Oxford handbook on the economics of networks*. Oxford: Oxford University Press.

Chapter 7
Spatial Inequality

Population and economic activities are not uniformly distributed across the territory, nor are cities of equal size, so the introduction of a geometric dimension to human activities opens up new aspects on problems of distribution. These heterogeneities do not necessarily represent problems of inequality, but they may do so indirectly because of the differences in physical environment, population and economic activities that they give rise to. Heterogeneities as far down as the local neighbourhood have proved to be important for individual life chances.[1] For this reason, it is important to understand the mechanisms behind them.

7.1 Allocation of Human Activities

The distribution of human activities across space has varied radically with both population size and cultural stage. The time paths of population distribution and inequality in well-being are different, but with interesting patterns of interaction. In some cases, concentration of population has facilitated exploitation and increased inequality; in others, stark inequalities have created incentives for migration. There are also general social forces affecting both settlement patterns and inequality.

A Short History of Location
Early human history was characterised by an extremely sparse population, which was relatively evenly distributed over a large territory. Movements in hunter-gatherer societies are guided by the need for access to natural resources such as water, food and material for shelter. The long transition to agricultural societies went via semi-sedentary stages, relying on improved methods for food storage, permanent housing, gradually concentrated land use and increasing capital accumulation.

---

[1]Chetty et al. (2014).

© The Author(s), under exclusive license to Springer Nature Switzerland AG 2022
P. Molander, *The Origins of Inequality*,
https://doi.org/10.1007/978-3-030-93189-6_7

Agriculture is highly dependent on water, and early civilisations logically developed around major rivers—Mesopotamia, the Nile Valley, the Indus Valley and others. The growth of these civilisations contributed to both an increasingly concentrated population and increasing inequalities.

With urbanisation followed increased division of labour, the emergence of administrative, religious and military elites, as well as further increases in inequality. As pointed out earlier (Sect. 5.1), the concentration of production possibilities in the riverine cultures facilitated exploitation of the majority by the ruling elites.

One of the main characteristics of agricultural society is that production and consumption for the overwhelming majority of the population are located at the same place. Industrialisation implied the first major unbundling in economic history—the separation between production and consumption.[2] After a transitory stage during which production to some extent remained decentralised within the putting-out system, it became concentrated in large industrial sites.

The location of industry depends on an intricate interplay between returns to scale, transport costs and demand. The consequence is that one and the same combination of such parameters sometimes can lead to more than one stable outcome, and that the current structure of settlement and industry will depend on history, a phenomenon referred to as *path dependence*.

Over time, transport costs have been reduced as a result of both spontaneous technical change and large infrastructure investments. Gradually, the trade-off between returns to scale and proximity to customers has therefore changed. New cities have been born. Within the industrial complex, branch concentration has increased.

A major second unbundling emerged as information technology reached a stage that facilitated diffusion of the production process across larger territories, even globally.[3] In recent decades, even certain types of qualified service production have been offshored to distant countries.

In parallel with these tendencies, the importance of human capital has increased and now represents the most important form of capital—4–5 times as important as physical capital in a typical OECD country.[4] This in turn has increased the importance of the system of education, in particular higher education. This form of capital follows a different allocation logic to traditional physical capital. Given that institutions of higher education are necessarily sparse, differences in access to higher education may be a source of regional inequalities, particularly when they are reinforced by positive feedback mechanisms in the accumulation of human capital.

Cities

In premodern history, the location of population centres was often determined by natural conditions. The importance of rivers was mentioned above, and the crossing of a river and trade route over land naturally developed into a settlement. Because of

---

[2] Baldwin (2006, 2016).

[3] Ibid.

[4] Liu (2011).

the importance of marine transports, natural harbours, river mouths or points where a river ceases to be navigable also easily developed into urban agglomerations.

But political decisions have also been important. In 1703, Tsar Peter decided to establish a major city on the boggy banks of the river Neva, a site where there was at the time only a fort just seized from Sweden and which was not obviously suited to urban development. Shortly after, St Petersburg was made the capital of Russia and is now the second city of the Russian Federation. During the twentieth century, similar political initiatives of moving a capital were taken in Turkey (Ankara), Brazil (Brasilia) and Kazakhstan (Astana, later renamed Nur-Sultan).

Because of the immense capital invested in a city, the status quo weighs heavily in the development path followed by an urban structure. Cities attract people and capital from outside, which defines a basic growth pattern. Auerbach seems to have been the first geographer to notice the regular pattern of city sizes, later generalised into Gibrat's law.[5] The pattern—that the size of cities starting with the smallest increases at a regular pace when plotted in a logarithmic diagram—has been reproduced with variations for a large number of countries. The outcome depends on the definition of cities, the degree of accuracy sought and other parameters.[6] By way of example, the size distribution of Swedish municipalities is shown in the diagram below.[7] The straight-line fit is far from perfect. The curve form and the drop of the curve towards the right-hand tail in this case indicates a combination of a lognormal distribution and a power law rather than a pure power law (Fig. 7.1).

Regional Inequality

A variable that is at least as relevant to policymaking as the size distribution of cities is the partition into urban and non-urban areas and how conditions of living differ between the two, or more generally, between central and peripheral regions of a country. Of particular interest is how large a fraction of the total population and economic activities is concentrated in the metropolitan areas. Among European countries, the fraction of the population living in the metropolitan region of the capital varies significantly, but the general tendency is that its share of the total population and GDP is currently increasing.[8]

A typical, long-term development scenario is illustrated in Fig. 7.2, which shows gross regional products in the 24 regions in Sweden from 1855 to 2007.

The general pattern is that differences between regions have been reduced during the one and a half centuries illustrated. This is clear from visual inspection of the diagram but is also verified by indicators such as the coefficient of variation or the Gini coefficient.[9] This is partly due to macroeconomic forces, partly to government

---

[5] Auerbach (1913).

[6] For surveys of the research debate with somewhat different perspectives, see Berry and Okulicz-Kozaryn (2012) and Arshad et al. (2018). See also Sect. 8.1.

[7] The Swedish system of public administration makes no distinction between cities and municipalities.

[8] Eurofound and European Commission Joint Research Centre (2019).

[9] Enflo et al. (2014).

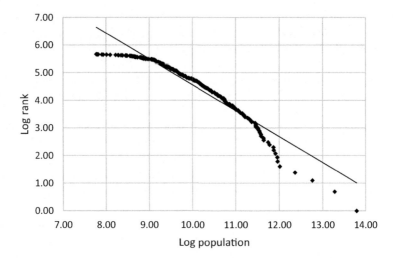

Fig. 7.1 Distribution of Swedish municipalities (logarithm of size against logarithm of ranking). The slope is −0.94. Source: Data from Statistics Sweden

measures, for instance the expansion of the system of education and a fairly ambitious system of equalisation within the public sector. The U-shaped curve of the metropolitan region should also be noted, indicating a recent concentration of economic activities to this region, albeit still less pronounced than at the beginning of the period.

The interplay between different factors of change gives rise to both centripetal and equalising forces. Productivity development in different sectors, infrastructure investments, and domestic and international migration have dominated during different historical periods. If the regional GDP per capita is used as an indicator of welfare, it can be expressed as the product of the following three factors:

• fraction of population in working age
• labour force participation within that fraction
• labour productivity.

Defining and comparing regional inequality across countries is difficult, for several reasons. Administrative structures differ, and the level of inequality as measured for instance by the Gini coefficient depends on the level of aggregation.[10] Further, depending on what is driving regional convergence or divergence, it is conceivable that regional and individual inequalities follow different paths.

As a concrete example, consider the development of regional and individual inequality in Sweden from 1860 to 2010.[11] During the first half-century, the country experienced rapid industrialisation, although the agricultural economy was still

---

[10]For a discussion of methodological problems, see Spiezia (2003) with further references.

[11]Sources of the account that follows are Enflo et al. (2014) and Enflo and Roses (2015).

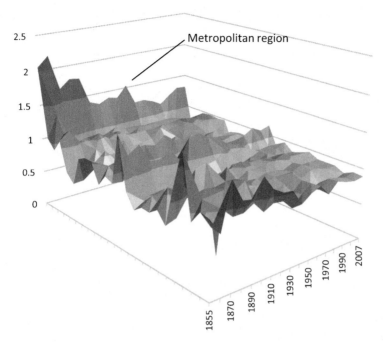

Fig. 7.2 Gross regional products in Sweden 1855–2007 (24 regions, metropolitan region at the far end). The value on the vertical axis represents the level of per-capita GRP in relation to the national average. Source: Data from Enflo et al. (2014)

dominant. From 1910 to 1940, an increasing share of the working population moved from the agricultural sector to industry, and the share of urban population passed the threshold of 50 per cent. But differences within sectors also increased, contributing to regional divergence. After the Second World War, labour movements from low-productive sectors contributed to a strong decrease in regional inequality up to 1980, where after inequality again started to increase, mainly as a result of increasing productivity differences between services and industrial production. A comparison between regional and individual inequality during the period studied is shown in Table 7.1.

Inequality between regions and individual inequality are obviously different matters. The current rapid increase in both regional and individual inequality stems from both productivity differentials between sectors and internal migration patterns. Detailed studies of urban growth[12] show that social interaction only explains about half of the scaling parameter of wage income and that differences in the sociodemographic composition of cities account for an important part. These differences are generated by selective migration of productive individuals into large

---

[12] Keuschnigg et al. (2019), Eliasson et al. (2020).

Table 7.1 Development of regional and individual inequality in Sweden 1860–present

Time period\Type of inequality	Regional	Individual
1860–1910	Decreasing	Increasing[a]
1910–1940	Increasing	Decreasing
1940–1980	Decreasing	Decreasing
1980–present	Increasing	Increasing

Source: Enflo et al. (2014), Enflo and Roses (2015) (regional); Roine and Waldenström (2009), Bengtsson et al. (2018) (individual)

[a] Individual inequality peaked towards the end of the period

cities—a form of domestic brain drain—which thus benefit disproportionately from the ongoing process of concentration.

7.2 The World System

Global, like national, inequality can be studied at the aggregate level or at lower levels, between regions or countries. Some of the problems and insights from regional analyses are reproduced, but with important differences.

History
During the last two centuries, the structure of inequality has undergone important changes. Global inequality—when the populations of all countries in the world are considered as one single population—rose during the nineteenth and twentieth centuries, peaked around the year 2000, and has since decreased marginally. Inequality between countries rose from a low value in 1820, peaked in the late twentieth century, then decreased somewhat in the present century. Figure 7.3 below shows the development in some more detail.

What strikes the eye in the above figure is the radical increase in between-country inequality. This pattern reflects what in the historical literature goes under the name of *the Rise of the West*—the process through which Europe and the United States rose from a marginal position at the beginning of the New Era to world dominance in the twentieth century.

In principle, the question of how to explain the Rise of the West could be addressed in the same way as the evolution of regional differences. Within a purely economic conceptual framework, one could study the global production system with reference to returns of scale, transport costs and demand patterns. We will return to this approach in Chap. 8, but here a more traditional, descriptive format will be used.

The concept of a *world system* is coupled to the name of Immanuel Wallerstein,[13] but many authors have tackled the problem of global patterns of power and economic and political dominance using a systems approach while not necessarily using this terminology. A large, traditionally oriented literature in history and economic

---

[13] See, for example, Wallerstein (2004).

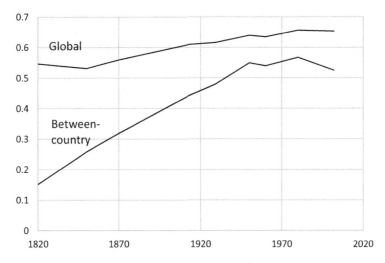

Fig. 7.3 World inequality 1820-present as illustrated by the global Gini coefficient and the Gini coefficient of between-country inequality. Source: Milanovic (2011), Table 2, with further references. Note: van Zanden et al. (2014) present somewhat lower figures for the global Gini coefficient during the nineteenth century. Figures for 2002 were later adjusted upwards to 0.599 and 0.766, respectively, based on new calculations of purchasing power parities (according to World Development Indicators)

history is devoted to the question of why the West took over and came to dominate the world.[14]

Ideational Approaches
A group of authors seek answers mainly in the history of mentality, religious or more general. An alleged *dynamism of Christianity* is an explanation that emanates from Christopher Dawson. A problem with this approach is the lack of dynamism in Europe while the Church dominated during the Middle Ages, while Islam turned out to be more dynamic during its expansion both westwards, through North Africa into the Iberian Peninsula, and eastwards, where it reached as far as Indonesia.

Value shifts initiating a more active and dynamic relationship have been proposed. Examples are English individualism (Macfarlane) and a taste for innovation (McCloskey). Against the former, one can object that individualist values can be traced both across civilisations and far back in history—the Gilgamesh epos, ancient Greek drama, and later for instance in Le Roy Ladurie's detailed mapping of the Cathar rebels in medieval France.[15] McCloskey's explanation appears to beg the question.

A specific *capacity for innovation* is the explanation advocated by David Landes and Lynn White, anchored in the history of technology. They refer to a number of

---

[14] For a survey, see Daly (2015), which contains references to the original works.
[15] Le Roy Ladurie (1975).

important innovations—water wheels, eye-glasses, printing and others—that were developed or perfected during the high Middle Ages. The problem here is that virtually all of these innovations existed in China or the Arab world centuries before and were imported, as shown by Joseph Needham and colleagues in their comprehensive work, *Science and Civilisation in China*.

Institutional explanations, linked to the ideational field but firmly within a modern social science tradition, can be found in authors such as Nathan Rosenberg, Douglass North and Richard Pipes, who in various combinations stress the importance of financial instruments, bookkeeping, property rights, and legal persons.

Technological Superiority

A second group of authors stress the importance of *technology*, particularly in the military field. While the Greeks produced advanced scientific results, later taken up and developed by the Arabs, Joel Mokyr stresses the link between science and technology—the quest for *useful* knowledge. In macroeconomics, the family of models of endogenous growth focus on this particular aspect of development.[16]

Rosenberg and Birdzell stress in particular the development of *marine technology* as such, whereas Cipolla points to its military use in the struggle for dominance at sea. Superiority in *military technology* in general is the main line of argument of William McNeill, Michael Roberts and Geoffrey Parker. The more or less permanent strife between small and large states in Europe provided a fertile testing ground for new technology, in contrast to relatively stable imperial China, where gunpowder and cannons were developed earlier. The Chinese explorer Zheng He carried cannons aboard his vessels when sailing the Indian Ocean in the early fifteenth century, but further explorations with a potential for colonisation were aborted by later governments.[17] In a sense, European political fragmentation became an asset in stimulating the development of military technology.

Material Explanations

Closer to economic modes of explanation can be found in a number of writers stressing natural endowments of various kinds. Jared Diamond and Alfred Crosby point to *basic geographic factors*, *flora* and *fauna*—a temperate climate, domesticable plants and mammals—but choose to consider the whole of Eurasia as one continent. In a similar vein, E.L. Jones has stressed that natural disasters and wars have on average been less lethal in Europe than in Asia, but on a more abstract level points to the general conditions for economic growth—state responsibility for infrastructure and other basic social functions, combined with reasonable economic freedom for citizens—which were also satisfied in China under the Song dynasty (960–1279). Jones thus invokes both geographic and institutional factors.

Imperialism and exploitation are recurrent themes in the literature. Combining economic and military aspects easily leads to the military-industrial complex formed

---

[16] See Sect. 8.2.

[17] Peterson (1994), Wade (2005).

by European states—England, the Netherlands, France, Spain, Portugal, to name the dominant ones—as an instrument to build worldwide empires.

These latter points provide common ground for a number of perspectives from economics and economic and political history. In the 1950s and 1960s, Raul Prebisch and Hans Singer chose to describe the world production system in core-periphery terms, similar to what the new economic geography decades later used in its analysis of regional economics and international trade.[18] The core, consisting of Europe and the United States, produced industrial goods for the whole world, while the periphery produced primary goods for the centre. As technology develops, there is a tendency for the centre to appropriate the gains because of the unequal distribution of power between the two parties. Prebisch stressed the decline in terms of trade of non-industrialised countries.

More recent empirical work in economic history has, broadly speaking, confirmed this picture. Beckert's detailed account of the cotton empire shows how the textile industry became the testing ground for prototypes of world capitalism instruments—the results of technological and financial innovation, but also including conquest of land, slave trade and military force.[19] The textile industry was not only a testing ground; for a long time, it was a dominant component of industry and the world trade system.

Traditional trade theory is based on comparative advantage and presents the case for free trade as essentially unproblematic. All parties to a trade exchange, even an apriori weaker party, are claimed to gain by specialising in the fields where they perform best, or least badly. In a broad summary of empirical research on the world trade system, Williamson has taken issue with this theory and identified the mechanisms by which the original balance between the West and the rest of the world tilted in favour of the West.[20] The free-trade regime had negative consequences for the East and South in three respects—deindustrialisation, inequality and volatility.

The extent of *deindustrialisation* is illustrated in Table 7.2.

In 1750, China and India each matched Europe in manufacturing output. Fifty years later, China still did, and more than half of world output came from China and India together. In 1913, the output of the developed core was 92.5 per cent of the total world manufacturing output, while China was down at 3.6 per cent and India at 1.4. The total manufacturing output grew during the period, although not enough to compensate for the reallocation illustrated in the table. The division of labour between the West and the East thus led to deindustrialisation and slowed down economic development in the East.

As for *inequality*, we know that stark inequalities are liable to lead to development traps that hinder economic growth.[21] Specialising in export of plantation-based agricultural production and mineral deposits increased inequality in the less

---

[18] Prebisch (1959), Singer (1964).

[19] Beckert (2014).

[20] Williamson (2011).

[21] See Sect. 5.1, Agricultural Societies.

Table 7.2 World manufacturing output 1750–1938 (per cent)

Year	India	China	Europe & the U.S.	Rest of the world
1750	24.5	32.8	27.0	15.7
1800	19.7	33.3	32.3	14.7
1830	17.6	29.8	39.5	13.3
1880	2.8	12.5	79.1	5.6
1913	1.4	3.6	92.5	2.5
1938	2.4	3.1	92.8	1.7

Source: Williamson (2011), Table 5.1, based on Simmons (1985) and Bairoch (1982)
Note: India refers to the entire subcontinent

industrialised countries because of their concentrated ownership, with negative effects on economic growth.

Volatility of revenues is another factor that is known to be negative for economic growth.[22] Higher volatility in raw materials and falling terms of trade slowed down economic growth in non-industrialised countries.

The *second unbundling* in Baldwin's terminology—the dissolution of the production process across the globe—can be viewed as a modern version of the above division of labour. It has proved to increase inequality in several respects.[23] Unbundling of production can lead to income divergence among identical countries (symmetry breaking). When countries are heterogeneous at the outset, top-bottom inequality is increased, and the world income distribution becomes more polarised; the income share of all developed and the most productive developing countries increases. Unbundling of production also increases within-country inequality in all countries. The direct effect of computerisation on inequality is zero without unbundling, but raises world inequality in a regime of unbundling. On the other hand, technology diffusion permits low-productivity countries to benefit from technological catch-up, leading to some income convergence in the long run.

The Political Dimension

To the problems associated with inequality can be added the difficulty of creating a well-functioning administration and reasonably democratic rule in a country characterised by stark inequalities. The risk of corruption is high, and integrity in the administration—a major determinant of state legitimacy—is difficult to achieve.[24] Economic power wielders will also resist the development of a functioning democracy, which is in any case a challenge in countries with weak or non-existent democratic traditions.[25]

---

[22] van der Ploeg and Poelhekke (2009).

[23] Basco and Mestieri (2019).

[24] Gilley (2006), Holmberg and Rothstein (2015), Rothstein (2021).

[25] Persson and Tabellini (2009), Lindenfors et al. (2020).

7.3 Summary

- Regional inequality follows a different logic to inequality between individuals or households. Growing individual inequality may coexist with decreasing regional inequality, and vice versa.
- Factors that work against the concentration of economic activities are natural resources, transport costs, costs of commuting (in time and money), land rents, and congestion costs. Centripetal forces are linkages between sectors (cluster effects), the attraction of markets (for labour and services), and positive external effects of knowledge. Actual patterns are affected by these conflicting forces, but prehistory as represented by the status quo of allocation of activities and population also weighs heavily.
- In principle, an economic approach can also be used in an analysis of international division of labour and trade patterns, but at this level, a number of other factors have intervened historically—military aggression, asymmetric economic relations, institutional differences, and others. The limited mobility of labour compared to that of capital is also important for the development of the power balance between them.

Bibliography

Arshad, S., et al. (2018). Zipf's law and city size distribution: A survey of the literature and future research agenda. *Physica A, 492*, 75–92.

Auerbach, F. (1913). Das Gesetz der Bevölkerungskonzentration. *Petermanns Geographische Mitteilungen, 59*, 74–76.

Bairoch, P. (1982). International industrialization levels from 1750 to 1980. *Journal of European Economic History, 11*, 269–333.

Baldwin, R. (2006). *Globalisation: The great unbundling(s)*. Prime Minister's Office/The Economic Council of Finland.

Baldwin, R. (2016). *The great convergence. Information technology and the new globalization*. Belknap Press.

Basco, S., & Mestieri, M. (2019). The world income distribution: The effects of international unbundling of production. *Journal of Economic Growth, 24*, 189–221.

Beckert, S. (2014). *The cotton empire. A new history of global capitalism*. Alfred E. Knopf.

Bengtsson, E., et al. (2018). Wealth inequality in Sweden, 1750–1900. *Economic History Review, 71*(3), 772–794.

Berry, B. J. L., & Okulicz-Kozaryn, A. (2012). The city size distribution debate: Resolution for US urban regions and megalopolitan areas. *Cities, 29*, Supplement 1, S17–S23.

Chetty, R., et al. (2014). Where is the land of opportunity? The geography of intergenerational mobility in the United States. *The Quarterly Journal of Economics, 129*(4), 1553–1623.

Daly, J. (2015). *Historians debate the rise of the west*. Routledge.

Eliasson, K., et al. (2020). Regional concentration of university graduates: The role of high school grades and parental background. *European Urban and Regional Studies, 27*(4), 398–414.

Enflo, K., & Roses, J. R. (2015). Coping with regional inequality in Sweden: Structural change, migrations, and policy, 1860–2000. *The Economic History Review, 68*(1), 191–217.

Enflo, K., et al. (2014). Swedish regional GDP 1855–2000: Estimations and general trends in the Swedish regional system C. *Research in Economic History, 30*, 47–89.

Eurofound and European Commission Joint Research Centre. (2019). European Jobs Monitor 2019: Shifts in the employment structure at regional level. *European Jobs Monitor series.* Luxembourg: Publications Office of the European Union.

Gilley, B. (2006). The determinants of state legitimacy: Results for 72 countries. *International Political Science Review, 27*(1), 47–71.

Holmberg, S., & Rothstein, B. (Eds.). (2015). *Good Government. The relevance of political science.* Edward Elgar.

Keuschnigg, M., et al. (2019). Urban scaling and the regional divide. *Science Advances* 2019(5), eaav0042 30 January 2019.

Le Roy Ladurie, J. (1975). *Montaillou, village occitan de 1294 à 1324.* Paris: Gallimard. Eng. trans. *Montaillou: Cathars and catholics in a French village 1294–1324.* London: Scolar.

Lindenfors, P., et al. (2020). The Matthew effect in political science: Head start and key reforms important for democratization. *Humanities and Social Sciences Communications, 7*, 106. https://doi.org/10.1057/s41599-020-00596-7

Liu, G. (2011). Measuring the stock of human capital for comparative analysis: An application of the lifetime income approach to selected countries. *Statistics directorate working paper* No. 41. Paris: OECD.

Milanovic, B. (2011). A short history of global inequality: The past two centuries. *Explorations in Economic History, 48*(4), 494–506.

Persson, T., & Tabellini, G. (2009). Democratic capital: The nexus of political and economic change. *American Economic Journal: Macroeconomics, 1*(2), 88–126.

Peterson, B. B. (1994). The Ming voyages of Cheng Ho (Zheng He), 1371–1433. *The Great Circle, 16*(1), 43–51.

Prebisch, R. (1959). Commercial policy in the underdeveloped countries. *American Economic Review, 49*(2), 251–273.

Roine, J., & Waldenström, D. (2009). Top incomes in Sweden over the twentieth century. In A. B. Atkinson & T. Piketty (Eds.), *Top incomes: A global perspective* (Vol. II). Oxford University Press.

Rothstein, B. (2021). *Controlling corruption. The social contract approach.* Oxford University Press.

Simmons, C. (1985). "Deindustrialization," industrialization, and the Indian economy, c. 1850–1947. *Modern Asian Studies, 19*, 539–622.

Singer, H. (1964). *International development: Growth and change.* McGraw Hill.

Spiezia, V. (2003). Measuring regional economies. *OECD Statistics Brief* 6, October 2003. Paris: OECD.

van der Ploeg, F., & Poelhekke, S. (2009). Volatility and the natural resource curse. *Oxford Economic Papers, 61*(4), 727–760.

van Zanden, J.-L., et al. (2014). The changing shape of global inequality, 1820–2000. *The Review of Income and Wealth, 60*(2), 279–297.

Wade, G. (2005). The Zheng He voyages: A reassessment. *Journal of the Malaysian Branch of the Royal Asiatic Society, 78*(1), 37–58.

Wallerstein, I. (2004). *World systems analysis. An introduction.* Duke University Press.

Williamson, J. (2011). *Trade and poverty. When the third world fell behind.* MIT Press.

Chapter 8
Spatial Allocation Models*

The theory of location of population and economic activities has a long history, counting such names as von Thünen, Christaller, and Lösch. A modern tradition has developed under the headings of *urban science* and *regional economics*. Inequality has not been in focus in this tradition, but regional inequalities nonetheless become relevant, since differences in regional development, hierarchies of cities and other forms of diversity affect living conditions.

Spatial models of human activities are variegated, even if attention is restricted to basically economic frames of reference. A direct link between standard economic analysis and regional development is *structural change*—the reallocation of capital and labour from less productive to more productive sectors. Because of inertia and barriers of various kinds, differences in growth rates and wages may persist for a very long time.

In the neoclassical model of *balanced growth*, there is a tendency towards convergence of investments and wages. In the developed version by Stiglitz, there is a possibility of divergence in growth under certain assumptions, as we have seen (Sect. 6.5). It is not difficult to imagine regional differences in savings rates or persistent heterogeneities in abilities, for instance, that would give rise to regional differences in a similar way.

In models of *endogenous growth*, productivity growth is determined by investments in education and research, leading to the possibility of growth differentials as soon as such investments differ across regions. There is also evidence of an internal brain drain between the regions in one and the same country.[1]

A novel approach to location problems, termed *new economic geography*, was launched by Krugman around 1980 and taken up by Fujita, Venables and Arthur.[2] This is a tradition that explicitly uses the concepts and methods advocated in the

---

[1] Keuschnigg et al. (2019), Eliasson et al. (2020).

[2] Krugman (1980, 1981), Fujita (1988), Venables (1996), Arthur (1994). A comprehensive presentation is given in Fujita et al. (1999).

present book—instability, symmetry breaking, branching points, and so on—and which also form the bulk of the present chapter. Only sketches of models will be given; for full proofs, the reader is referred to Fujita et al. (1999), with further references.

8.1   Cities and Regions

The building-blocks of modern location theory—production technology, transport costs, demographic change—can already be found in early contributions, but new mathematical tools and computers have permitted the development of more sophisticated analyses.

The first question to be asked is why and how cities emerge at all. The simple answer is that there are returns to scale, whether in administrative or economic matters, which require some form of concentration of capacities. The fact that urban cultures emerged independently in different parts of the world proves that cities are an answer to universal problems in the course of human social and cultural development. In consequence, it can be assumed that there exists at least one city, and that the analytical problems to be solved revolve around issues such as the number of cities, their relative size, possible hierarchies, and so on.

Basic Approach
The pattern of localisation that results from a set of feasible alternatives is a complex function of producer and consumer behaviour, transport costs and other parameters. A natural way to proceed is to start with highly simplifying assumptions and to increase realism by successively relaxing them. Because returns to scale are essential to the outcome, classical assumptions on decreasing returns are useless, and the central building-block is instead the Dixit-Stiglitz model of monopolistic competition.[3]

Products are divided into manufactured (M) and agricultural (A), where the manufacturing sector is characterised by a large number of varieties, and the agricultural sector is perfectly competitive. Consumer utility obeys a standard Cobb-Douglas assumption:

$$U = M^{\mu}A^{(1-\mu)}$$

The composite index M consists of quantities $m(i)$ of the different varieties, and M is defined as the constant-elasticity-of-substitution function of the $m(i)$:

---

[3] Dixit and Stiglitz (1977).

$$M = \left[\int m(i)^{\rho} di \right]^{1/\rho}, 0 < \rho < 1.$$

The integral is taken over the available range of varieties of manufactured products. The parameter ρ is a measure of the preference for variety among these products. Alternatively, one can define $\sigma > 1$ via

$$\sigma = 1/(1 - \rho),$$

in which case σ is the elasticity of substitution between any two of the varieties.

An important restriction, which is necessary to prevent the manufacturing industry from collapsing into one point, is that ρ be greater than μ:

$$\rho > \mu.$$

This is referred to as the *no-black-hole condition*.

A composite price index G can be derived from the condition of maximisation of utility. If the range of products is increased, competition is intensified, and prices (G) will fall.

Together with the price of agricultural products (p^A) and total income Y, this leads to the maximisation problem

$$max\ U = M^{\mu} A^{(1-\mu)} subject\ to\ GM + p^A A = Y.$$

The range of manufactured products offered is determined endogenously on the basis of profit maximisation. Important parameters are T^M_{rs} and T^A_{rs}, which model transport costs from location r to location s for manufactured and agricultural products, respectively. The value of a product is reduced by the factor in question, when the product is transported from r to s. An index of the trade cost, or friction, is Z defined by

$$Z = \left(1 - T^{(1-\sigma)} \right) / \left(1 + T^{(1-\sigma)} \right).$$

The number Z varies between 0 and 1, where 0 corresponds to zero trade costs and 1 to no trade.

If the situation is limited to the simplest case of two regions, it is interesting to see what characterises the symmetric solution in which labour input L and income Y are the same in the two regions. If the nominal wage rate is denoted by w, one gets

$$\frac{dY}{Y} = Z \frac{dL}{L} + [\% + Z(1 - \sigma)] \frac{dw}{w}$$

The interpretation of this equation is that an increase demand for manufactured goods will result partly in increased employment, partly in higher wages, where the proportions are determined by the trade friction parameter Z. In consequence, regions with a higher demand for manufactured goods will enjoy higher nominal wages. But an increase in L will result in a lower price index, so real wages will also be higher.

In order to introduce some dynamics into the model, a number of simplifying assumptions permit one to concentrate on workers in the manufacturing industry. As before, let the different regions be denoted by r, and let λ_r be the fraction of workers in region r. If the real wage in region r is denoted by ω_r, the average real wage is given by

$$\overline{\omega} = \sum_r \lambda_r \omega_r$$

Workers are assumed to be attracted to a region in proportion to the difference between the regional real wage and the average real wage:

$$\dot{\lambda}_r = \gamma(\omega_r - \overline{\omega})\,\lambda_r \tag{8.1}$$

This will be recognised as the replicator equation from the framework of evolutionary game theory (Sect. 2.4). Successful regions are so-to-say rewarded with an inflow of labour.

In order to determine the patterns of production and the resulting wage differentials that are compatible with these assumptions, the stationary points of Eq. (8.1) need to be studied. The decisive parameter is the transport cost. In the case of two regions, some typical outcomes are shown in Fig. 8.1. The horizontal axis represents the fraction of workers in manufacturing industries in one of the regions, varying between 0 and 1. The vertical axis represents the wage differentials between the two regions.

The pattern of location as transport costs are successively reduced—the secular change experienced in most parts of the world—can be deduced from the stability or instability of the equilibria shown in the figure, marked by circles. If transport costs are high, there is only one equilibrium. This is symmetric—wages are equal in the two regions, and manufacturing industry is divided equally—and it is stable, as indicated by the arrows.

As transport costs are reduced to an intermediate level, five equilibria emerge. Three of these are stable: the previous symmetric equilibrium and two equilibria at the edges, in which the entire manufacturing industry is concentrated into one region. Between these equilibria, there are two unstable ones, asymmetric with respect to shares in manufacturing but with equal wages. Because of their instability, they will not survive, however, but develop into one of the adjacent stable equilibria.

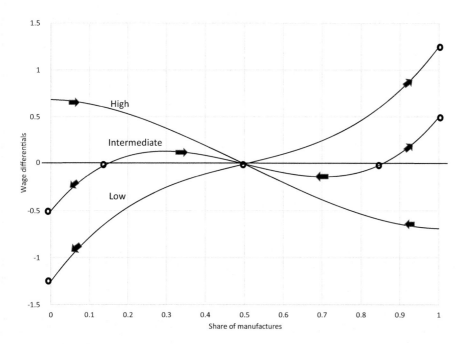

Fig. 8.1 Partition of the labour force (horizontal axis) and wage differentials between the two regions (vertical axis) for the three different levels of transport costs (high, intermediate and low). Source: General background from Fujita et al. (1999)

The stable equilibria are potential solutions of the allocation problem. Whether they will be exploited or not depends on the parameters of the problem and in some cases on prehistory.

When transport costs are further reduced, the symmetric equilibrium at the centre collapses, and only the asymmetric solutions at the edges survive.

In summary, the model illustrates how qualitative changes may occur at certain critical values of a driving parameter, in this case the transport cost. These *bifurcation points* have somewhat different characteristics depending on the patterns of interaction between the components of the system.[4] An interesting property is that if transport costs were, for some reason, to rise again, the system would not follow the same path as when costs decreased. The existing structure is a form of memory of the past, called *hysteresis* in technical language, or *path dependence*.

Although highly suggestive, this model is oversimplified in important respects, and conclusions should not be stretched too far. Generalising the description from two regions to three or more reproduces some of the previous patterns but in general requires partly new analytical tools.

---

[4] The mathematical discipline devoted to the study of such critical points is *catastrophe theory*. This field appears to have lost some of its attraction in recent decades; for a discussion, see Rosser (2007).

New Cities, Hierarchies and Size Distribution

Von Thünen's original model of the interaction between the city and the surrounding hinterland can be developed into an analysis of the formation of new cities. For this purpose, one can imagine a one-dimensional space along which agricultural and manufacturing activities take place. Transport costs and other production parameters are assumed given in this context, and the driving force is *population growth*.

The trick in this case is to define a market potential function, $\Omega(r)$, where r is the distance to the centre:

$$\Omega(r) = \frac{\omega^M(r)^\sigma}{\omega^A(r)^\sigma}$$

Here, $\omega^A(r)$ is the wage rate in the agricultural sector, $\omega^M(r)$ is the (potential) wage rate that the manufacturing industry can offer at distance r from the city, and σ as before is the elasticity of substitution between manufactured goods. A structure with a single city is stable as long as the manufacturing industry cannot offer higher wages at location r than the agricultural sector, that is, as long as $\Omega(r) \leq 1$. When this condition fails to be satisfied, two new urban centres are born symmetrically with respect to the original city. The adjustment dynamics are complex, depending on the basic parameters of the system. When population continues to grow, new cities are born at a regular pace, being serviced from symmetrically located agricultural areas.

What is typical of urban and regional structures around the world is that living conditions differ. Large cities not only have a larger population but also a more diversified industry, a broader offer of educational possibilities and a richer cultural life. There is in other words a *hierarchy* between cities that a theory should preferably be able to reproduce. The above framework does permit such an extension. What is needed in this context is that industries be allowed to differ with respect to their elasticities (σ- or ρ-values) and transport costs. An industry M_i is said to be of a *higher order* than M_j if it has a lower ρ-value (higher elasticity of substitution), or alternatively has the same ρ-value but a lower transport cost.

If the previous analysis is reiterated using this differentiated description, it can be shown that new cities are born following a regular pattern. Low-order industries are the first to emerge outside the centre; eventually, higher-order industries follow suit.[5] The highest-order city always provides the full gamut of industries, and lower-order cities can be ranked according to the number of manufacturing categories present.

Size Distribution of Cities

That the *size of cities* in a country tends to follow a power law distribution is an old observation, as pointed out in the previous chapter.[6] The number of cities that are bigger than a certain size S tends to vary proportionally to $S^{-\alpha}$, for some α close to 1. The accuracy of this approximation varies, somewhat depending on the definition

---

[5] See Fujita et al. (1999), Fig. 11.3, for an example.
[6] See Sect. 7.1, Cities.

of a city, and some authors claim that a lognormal distribution is more justified.[7] A number of explanations have been put forward to account for this regularity. It is a regularity, not a natural law, and in any case, the outcome of a ranking will in general depend on the definition of a city and its boundaries. In any case, a satisfactory approach to the birth and growth of cities requires a more diversified taxonomy than just dividing activities into agriculture and manufacturing.

Building on the general micro-foundation approach sketched above, Hsu has shown that the two problems of city hierarchies and size distribution can be approached simultaneously.[8] The driving force behind the city size differences and the emerging hierarchy is the heterogeneity in economies of scale across industries. In fact, the rank-size regularities that have been observed for cities also holds for the presence of industries, as shown previously by Mori et al.[9] The central place hierarchy generates the skewed city size distribution without any other assumption than a mathematical restriction satisfied by common statistical distributions.[10] The central place hierarchy can be shown to exhibit a fractal structure, that is, the relationship between a level and its sublevels is constant across levels.

8.2   International Division of Labour

In the above analysis, no assumptions have been made on political boundaries. In principle, regional or national location theory should be applicable in a global setting, making models of international allocation and trade essentially regional economics writ large. It should not come as a surprise that many results do carry over to the global level.

From Regions to Countries

There are important differences between regions and countries, nonetheless. At the international level, there may be politically created obstacles such as tariffs or import quotas, but there are also natural impediments such as language barriers and cultural differences, and social obstacles not directly linked to trade policy, such as differences in technical standards and limited portability of social security rights. A simple way of modelling some of these barriers is to assume that labour is immobile. The role played in the levelling of production conditions is taken over by intermediate goods.[11]

In a world where labour cannot migrate, agglomerations cannot emerge as in the intranational process. Instead, agglomerations are formed by industries that exploit

---

[7] Eeckhout (2004).

[8] Hsu (2011).

[9] Mori et al. (2008), based on Japanese data. The slope in this case is around -0.7.

[10] The distribution of scale economies should be *regularly varying*; see Hsu (2011), p. 914.

[11] Fujita et al. (1999), Chap. 14.

the benefits of linkages between various stages in the production process. The driving force is first assumed to be *diminishing trade costs*. As in the previously analysed case, qualitative change can be triggered by small variations of exogenous parameters. Asymmetric patterns of industrialisation may emerge as different countries specialise in different product clusters, and such differences also give rise to wage differentials and differences in standards of living, now that the mobility of labour is limited.

It is also possible to relate the pattern generated to global development in recent decades, during which some previous agricultural economies have made the transition to partly industrialised countries, whereas others have remained in a state of low or highly volatile growth, being largely dependent on raw material exports.

An alternative driving force against which the model can be tested is general growth, expressed as a steadily *rising demand for manufactured goods*. Technically, this can be formulated as a constant demand for agricultural goods, while the share of the consumption function that captures demand for manufactured goods is increasing over time.

When this analysis is extended to the case of many products and many countries, a pattern of successive industrialisation waves can be deduced. The most labour-intensive industries are the first to migrate from the industrial economies, starting a process of wage equalisation between the countries involved. In successive waves, new countries are integrated. Later entries are facilitated by the linkages created with already established industries. The previous pattern of relatively rapid regime changes is reproduced, as well as the grouping of countries at different levels of living standards.

Models of Endogenous Growth

A different approach to the gap between growth rates in different countries is represented by models of endogenous growth, focusing on the accumulation of knowledge rather than decreasing transport costs. Early contributions were made by Romer and Lucas.[12] The motivation for this approach is partly the same as that which lies behind the new economic geography—that countries differ in their growth rates, and that these differences have tended to grow during long periods instead of converging, as certain traditional models would suggest. Romer also notes that the growth rate of the United States economy tended to increase from the early nineteenth to the late twentieth century[13]—an acceleration that is difficult to reproduce in a classical balanced-growth model.

The decisive difference from the classical tradition is acknowledging the existence of *increasing returns to scale*. The production function has knowledge as an essential factor of production and exhibits increasing returns to scale both in this and other factors. *Knowledge* also has a *collective dimension*, in the sense that the collected knowledge in all producers enters as a factor in the production function

---

[12]Romer (1986, 1990), Lucas (1988).

[13]Romer (1986), Table 2.

of each producer. This reflects the fact that knowledge cannot be kept fully private but is to some extent disseminated in society.

What keeps growth within certain bounds is that knowledge production itself exhibits decreasing returns to scale in its inputs. Together with other technical assumptions, this ensures the existence of a solution to the profit maximisation problem. The social optimum cannot be supported as a competitive equilibrium in the absence of government intervention, however. The reason is precisely the collective character of the socially available knowledge. Producers will take this asset as given and will not invest at the socially optimal level of knowledge production, even if it expects other firms to do so. Instead, each producer will produce less than optimally, reflecting the classical collective-action problem.

Nonetheless, there exists a competitive, suboptimal equilibrium solution, in which both capital and consumption grow without bounds. In a multi-country setting, in which each country is modelled as a closed economy (no trade), there will be no tendency towards convergence in the level of per capita production. Even if all countries start out with the same stock of knowledge, small disturbances will cause differences to grow over time. In fact, the assumption of fully closed economies is not necessary for this conclusion.[14]

In a more recent paper, Jones and Romer contrast the classical approach as exemplified by Kaldor's theory of capital accumulation with the modern theory of endogenous growth.[15] Half a century after the post-war period of recovery, the world economy looks very different. Markets have extended, and the flows of goods and services, knowledge, and finance have increased radically. Growth has accelerated, but growth rates vary substantially across the world, resulting in large gaps in productivity and income. Human capital per worker has also risen throughout the world, but the price of human capital has not fallen in spite of rising supply. The authors relate these facts to the interaction between ideas, institutions, people and human capital—factors that did not enter the analysis in the older generation of models.

The Effects of Financial Integration

The description of the economy in the above models is material, limited to labour, physical investments and transport. Financial integration has been an important component of global integration in recent decades, however: the total daily exchange in financial markets increased tenfold from the late 1980s to the late 2010s.[16] A natural question to ask in this context is how this affects development prospects and inequality.

Traditionally, it has been more or less taken for granted that financial integration benefits the parties involved in the same ways as free trade has been assumed to do.[17]

---

[14] Romer (1986), p. 1033.

[15] Jones and Romer (2010).

[16] Statistics Sweden.

[17] See references in Sect. 5.6.

But much as economic historians have problematised trade exchange and international division of labour by pointing to de-industrialisation, inequality and volatility, financial experts have begun to raise questions about the benefits of financial integration. The financial crises of the late 1990s and 2008 onwards have intensified this discussion.

Like most other decision problems, this is a question of benefits and costs. In terms of gains, Coeurdacier et al. (2020) have shown them to be modest. Further, most studies have assumed that the growth effect of integration is the same everywhere, but it is necessary to recognise the heterogeneity among the partners involved, depending on capital scarcity, risk and size of the country. This heterogeneity explains why it has been difficult to establish results that are robust across time and space. Safer, that is, developed countries benefit more from financial integration than riskier countries—emerging market economies. There is consequently a risk that financial integration, compared to autarky, reduces the rate of growth in developing and emerging economies if their level of risk is high compared to developed countries.

These ambiguous effects of integration, particularly the substantial volatility of financial flows, creates problems for financial policy management in emerging economies.[18] There are also indications that a floating-currency regime does not have the protective effects previously assumed.[19]

But conflicts between financial integration and growth may go even deeper. In a stylised model, Matsuyama has shown that inequality between countries may develop as a result of financial integration, starting from an autarkic regime.[20] The world economy consists of identical countries differing only in their levels of capital stock. An integration of financial markets affects the balance between equalisation and divergence tendencies. In the autarkic regime, the world economy has a unique stable steady state, which is symmetric. When the financial markets are integrated, symmetry is broken under certain conditions. The previous steady state may become unstable, and the emerging steady states will be unequal. Whether this separation into rich and poor countries will occur or not depends on a parameter λ, $0 < \lambda < 1$, that summarises the degree of imperfection of capital markets. When $\lambda = 0$, all projects have to be self-financed; when $\lambda = 1$, there are no borrowing constraints.

The separation of countries into rich and poor is not Pareto-sanctioned; the rich are richer and the poor are poorer in the asymmetric stable states than in the symmetric steady state. Aggregate global wealth also diminishes. The explanation for this outcome is that countries living in an integrated global financial market have to compete for the same financing. Given that countries are always hit by disturbances that are heterogeneous, countries that are unlucky will experience a downward spiral of lower profitability and increasing difficulties of financing, whereas countries that are lucky face the opposite prospect.

---

[18] Ostry et al. (2010), Ghosh et al. (2017).

[19] Passari and Rey (2015), Rey (2018).

[20] Matsuyama (2002, 2004).

Some important restrictions of the analysis are that only the index of financial market imperfection is assumed to be unaffected by globalisation, and that only the extreme cases—autarky and full globalisation—are considered. Various generalisations are possible. Zhang has shown that a certain amount of wealth inequality reduces the risk of instability of the symmetric equilibrium.[21]

Poverty Traps
A *poverty trap* is a stable equilibrium of an economy in which aggregate welfare is lower than it could be, given the general conditions of natural resources, climate, etcetera. The origin of such traps is some form of external effect—individual decisions depending on the decisions of others in one way or another. We have already encountered a trap mechanism in the formation of individual human capital (Sect. 3.3) and, at the collective level, in agricultural economies (Sect. 5.1) and in models of herd behaviour and segregation (Sect. 6.3). In the present context, the aim is to show how entire countries may be unable to escape from an equilibrium even if it is dysfunctional and large groups in society realise this fact.

A range of different mechanisms are possible.[22] Becker et al. analyse *fertility choices* as a function of the availability of human capital.[23] In a developed state, rates of return on human capital investments are high relative to rates of return on more children, whereas the opposite is true in a poor state. This leads to two stable equilibria—one with large families and little human capital, the other with small families and the prospect of growing human capital.

Basu analysed *child labour* in a similar vein, against the backdrop of a large empirical literature.[24]

Social norms and expectations, for instance when it comes to honesty or corruption, exhibit dynamics similar to the herd behaviour examined in Sect. 6.3. Tirole applied this approach to the problem of *corruption*.[25]

Within the economic sphere, the above model by Matsuyama shows how a group of countries can be caught in a vicious circle of *low profitability and low investments*. A fundamental observation made already by Adam Smith is that the degree of division of labour is limited by the size of the market. Advanced production requires highly specialised suppliers of intermediates, and it may be difficult to create the size and level necessary for the production of the final goods or services. The industrial clusters described earlier in this chapter illustrate this phenomenon.

The general form of the mechanism is illustrated in Fig. 8.2.

An individual, a group or an entire economy faces a decision, which in general terms can be called an investment. The horizontal axis corresponds to the current state, while the vertical axis refers to future states that result from current

---

[21] Zhang (2017).

[22] A broad overview is supplied in Banerjee and Duflo (2011).

[23] Becker et al. (1990).

[24] Basu (1999).

[25] Tirole (1996).

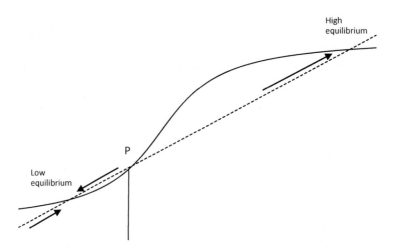

Fig. 8.2 The poverty trap mechanism. Motions starting to the left of the point P will converge to the low equilibrium, while motions starting to the right of P will converge to the high equilibrium

investments. The curve describes the relationship between current investment decisions and future states. Possible equilibria are points where current investments exactly meet what is necessary to support the current state (intersections with the straight line). For stability, the straight line must intersect the curve from below. In order to escape from lower equilibrium, the agents must somehow pass the critical point P.

A number of policies for escaping from poverty traps have been suggested. Rosenstein-Rodan's argument for a big push, written during the Second World War and dealing with the problems of eastern and south-eastern Europe, is a classical reference.[26] There is a large literature related to the debate in the 1990s about the so-called "Washington Consensus".[27] A general, empirically based conclusion is that the case for state intervention is stronger than claimed by the Washington Consensus. The experience of the Asian countries that made the transition from the low-income to the middle-income category is particularly relevant.[28]

8.3   Summary

• Most classical models of economic growth and development exhibit spatial inequalities when a spatial dimension is added.

---

[26] Rosenstein-Rodan (1943).

[27] Williamson (1990).

[28] See Wade (2004) for a review.

- Combining returns to scale, transport cost and standard assumptions on consumer and producer behaviour, it is possible to reproduce empirically observed urban and regional structures, such as the genesis of cities and urban hierarchies. What level of inequality develops as a result of this process depends on the assumptions made. Differences in conditions of living may emerge even when labour is assumed to have the possibility of migrating.
- *Symmetry breaking* is a recurrent phenomenon, implying that small changes in external parameters (e.g., transport costs or population size) can give rise to qualitative changes in the aggregate structure. Large differences may develop between cities or regions with essentially the same basic conditions for growth and development.
- International division of labour can be modelled under the assumption that labour is immobile. Many of the results from the regional analysis carry over to the international level, with the difference that it is the industry, rather than labour, that migrates in order to form more or less advanced clusters.
- Models of endogenous growth, where knowledge enters as a central factor of production, can explain some of the divergences in growth rates that have been observed between regions and countries. The collective character of knowledge will lead to a suboptimal level of the production of knowledge in society in the absence of government interventions.
- Financial integration has an ambiguous effect on development and inequality. Both theoretical and empirical analyses indicate that integration can magnify differences between developing and developed countries.
- *Poverty trap* is a catchphrase for clusters of problems experienced by many developing countries, leading to low- or no-growth social equilibria. Due to complex interactions between different policy areas, poverty traps represent difficult challenges that may require coordinated efforts in a number of policy areas.

Bibliography

Arthur, W. B. (1994). *Increasing returns and path dependence in the economy*. The University of Michigan Press.

Banerjee, A., & Duflo, E. (2011). *Poor economics. A radical rethinking of the way to fight global poverty*. Public Affairs.

Basu, K. (1999). Child labor: Cause, consequence, and cure, with remarks on international labor standards. *Journal of Economic Literature, XXXVII* (September 1999), 1083–1119.

Becker, G. S., et al. (1990). Human capital, fertility, and economic growth. *Journal of Political Economy, 98*(5), Part 2, S12–S37.

Coeurdacier, N., et al. (2020). Financial integration and growth in a risky world. *Journal of Monetary Economics, 112*, 1–21.

Dixit, A. K., & Stiglitz, J. E. (1977). Monopolistic competition and optimum product diversity. *The American Economic Review, 67*(3), 297–308.

Eeckhout, J. (2004). Gibrat's law for (all) cities. *American Economic Review, 94*(5), 1429–1451.

Eliasson, K., et al. (2020). Regional concentration of university graduates: The role of high school grades and parental background. *European Urban and Regional Studies, 27*(4), 398–414.

Fujita, M. (1988). A monopolistic competition model of spatial agglomeration: Differentiated product approach. *Regional Science and Urban Economics, 18*(1), 87–124.

Fujita, M., et al. (1999). *The spatial economy. Cities, regions, and international trade.* Cambridge, MA: MIT Press.

Ghosh, A. R., et al. (2017). *Taming the tides of capital flows. A policy guide.* MIT Press.

Hsu, W.-T. (2011). Central place theory and city size distribution. *The Economic Journal, 122*(563), 903–932.

Jones, C. I., & Romer, P. M. (2010). The new Kaldor facts: Ideas, institutions, population, and human capital. *American Economic Journal: Macroeconomics, 2*(1), 224–245.

Keuschnigg, M., et al. (2019). Urban scaling and the regional divide. *Science Advances* 2019(5), eaav0042 30 January 2019.

Krugman, P. R. (1980). Scale economies, product differentiation, and the pattern of trade. *American Economic Review, 70,* 950–959.

Krugman, P. R. (1981). Trade, accumulation, and uneven development. *Journal of Development Economics, 8*(2), 149–161.

Lucas, R. E. (1988). On the mechanics of economic development. *Journal of Monetary Economics, 22,* 3–42.

Matsuyama, K. (2002). Explaining diversity: Symmetry-breaking in complementarity games. *AEA Papers and Proceedings, 92*(2), 241–246.

Matsuyama, K. (2004). Financial market globalization, symmetry-breaking, and endogenous inequality of nations. *Econometrica, 72,* 853–884.

Mori, T., et al. (2008). The number-average size rule: A new empirical relationship between industrial location and city size. *Journal of Regional Science, 48*(1), 165–211.

Ostry, J., et al. (2010). Capital inflows: The role of controls. *IMF Staff Position Note* SPN/10/04 (February 19, 2010). Washington, DC: The International Monetary Fund.

Passari, E., & Rey, H. (2015). Financial flows and the international monetary system. *The Economic Journal, 125,* 675–698.

Rey, H. (2018). Dilemma not trilemma: The global financial cycle and monetary policy independence. *NBER working paper* No. 21162. Cambridge, MA: National Bureau of Economic Research.

Romer, P. M. (1986). Increasing returns and long-run growth. *The Journal of Political Economy, 94*(5), 1002–1037.

Romer, P. M. (1990). Endogenous technological change. *Journal of Political Economy, 98*(5), Part 2, S71–S102.

Rosenstein-Rodan, P. N. (1943). Problems of industrialisation of Eastern and South-Eastern Europe. *The Economic Journal, 53*(210/211), 202–211.

Rosser, J. B. (2007). The rise and fall of catastrophe theory applications in economics: Was the baby thrown out with the bathwater? *Journal of Economic Dynamics & Control, 31,* 3255–3280.

Tirole, J. (1996). A theory of collective reputations (with applications to the persistence of corruption and to firm quality). *The Review of Economic Studies, 63*(1), 1–22.

Venables, A. J. (1996). Equilibrium locations of vertically linked industries. *International Economic Review, 37*(2), 341–359.

Wade, R. (2004). *Governing the market. Economic theory and the role of government in East Asian industrialization.* New edition of the original 1990 book. Princeton, NJ: Princeton University Press.

Williamson, J. (1990). What Washington Means by Policy Reform. In J. Williamson (Ed.), *Latin American readjustment: How much has happened?* Washington, DC: Institute for International Economics.

Zhang, H. (2017). Wealth inequality and financial development: Revisiting the symmetry breaking mechanism. *Economic Theory, 63,* 997–1025.

Chapter 9
Philosophical and Political Considerations

In an often-quoted passage from *A Treatise on Human Nature*, David Hume remarked[1]:

> In every system of morality, which I have hitherto met with, I have always remark'd, that the author proceeds for some time in the ordinary way of reasoning, and establishes the being of a God, or makes observations concerning human affairs; when of a sudden I am surpriz'd to find, that instead of the usual copulations of propositions, *is*, and *is not*, I meet with no proposition that is not connected with an *ought*, or an *ought not*. This change is imperceptible; but is, however, of the last consequence. For as this *ought*, or *ought not*, expresses some new relation or affirmation, 'tis necessary that it shou'd be observ'd and explain'd; and at the same time that a reason should be given, for what seems altogether inconceivable, how this new relation can be a deduction from others, which are entirely different from it. But as authors do not commonly use this precaution, I shall presume to recommend it to the readers; and am persuaded, that this small attention wou'd subvert all the vulgar systems of morality, and let us see, that the distinction of vice and virtue is not founded merely on the relations of objects, nor is perceiv'd by reason.

Hume remarked that his observation might be "of some importance", which is an understatement, given the voluminous literature that has been devoted to the Is-Ought problem since. Hume's statement has been both supported and criticised during the almost three centuries that have passed since its publication. Gerhard Schurz made a decisive contribution to the discussion in the late 1990s with his book *The Is-Ought Problem*.[2] Schurz uses the full power of modern logic to conclude that Hume was essentially right, with some qualifications. One cannot go from an *is* (factual statement) to an *ought* (ethical or political recommendation) without the aid of some connecting link, what Schurz calls *analytical bridging principles*.

That Hume was essentially right does not necessarily imply that recommendations are totally independent of facts. If they were, the battle over facts in major policy areas concerning, for instance, climate change or links between smoking and cancer would

---

[1] Hume (1739–40), Book III, Part I, Section I, last passage.
[2] Schurz (1997).

be incomprehensible. These debates have been heated precisely because facts are relevant to policy decisions. A simple example may illustrate the relationship. Consider a group of visitors to some famous building being informed by a local guide. "This house was completed in 1724," the guide explains. This is a factual statement that in normal cases will not release any reflexes to act among the visitors. Immediately after this statement, a janitor rushes into the room, yelling "The house is on fire!" This is also a factual statement, but unlike the previous one, it will release a reflex, in this case to dash for the emergency exit. The difference is, of course, that the necessary bridging principle is immediate—the visitors want to survive.

In conclusion, Hume's statement is logically correct but practically and politically less interesting. Facts are highly relevant to policy decisions. In matters of public policy, both general knowledge from the social sciences and information specific to the problem at hand must be collected and organised. False statements should be identified and discarded, highly probable statements distinguished from unlikely ones, and so on. Obviously inferior alternatives should be eliminated. If two alternatives for achieving a given political goal are considered and it can be shown that one of them is less costly while achieving the same goal, it is difficult to argue that the other should be implemented.

The present chapter aims to draw some general policy conclusions related to inequality from the research synthesised in previous chapters. In the first section, the most important stylised facts are summarised as a platform for the ethical and policy-related discussions in the following sections.

9.1 Taking Stock

Chapters 3 and 4 analysed the mechanisms behind inequality in an individual perspective, Chaps. 5 and 6 the same question in a social context, and Chaps. 7 and 8 added a spatial dimension. Obviously, these three approaches are not completely distinct. Individual development is unthinkable without a social context, and both individual and societal processes necessarily take place somewhere in three-dimensional space. But the categorisation is nonetheless helpful when sorting the mechanisms involved, for the purpose of drawing policy conclusions. We start with the individual perspective.

Individual Differences Tend to Grow Over the Life Course

There is general consensus that body height distribution in modern healthy populations is nearly symmetrical, well approximated by a normal distribution with a moderate standard deviation. This has not always been the case. Historically, the distribution has been left-skewed due to malnutrition, that is, an adverse environment that does not permit the individual to develop its full potential.[3] Further,

---

[3] See Staub et al. (2015) for an example (Swiss data).

body height is strongly genetically determined. Within Europe, heritability in Caucasian populations is strong, estimated to be around 0.9, with a somewhat lower figure for women than for men.[4]

By contrast, the heritability of cognitive skills has been estimated to lie between 0.07 and 0.13, using similar methods.[5] The span for social and environmental influence is obviously much wider in this case than in the case of body height.

This insight, although not always made explicit, is reflected in the predominant use of *the capital metaphor* in thinking and research about individual life course development. This holds for a variety of models of human capital formation and development, whether designating cognitive skills only or comprising also non-cognitive skills and health variables.

Capital typically grows proportionally, that is, differences at the outset are magnified in absolute terms, increasing the spread and introducing a skewness to the right. If social capital defined on a network basis is added, the tendency towards increasing inequality will be even stronger.

This is the fundamental mechanism behind inequality at the individual level.

Dynamic Complementarity

Success begets success, and not only within one and the same dimension of human capital. High performance in school strengthens self-confidence and may improve health. Conversely, somatic or psychic ill-health is likely to impair performance in education or working life. The negative consequence of this interdependence is the existence of vicious circles in individual life courses. The positive side is that investments that strengthen the individual's capacity in one field may have positive side effects in other areas.

The Social Component in Intergenerational Transmission Dominates

This statement is more or less a corollary to the preceding one. Physical, and to some extent social, capital is easily inherited, and empirical research shows that societies where physical capital is important have a stronger coefficient of transmission and, as a result, higher levels of inequality.

An aspect of particular importance to the lower part of the distribution of resources concerns the exposure to risks and the capacity to cope with them. Children and young persons with a weak socioeconomic background are, broadly speaking, more exposed to hazards of various kinds—environmental, health-related, and so on. They also experience more adverse effects when exposed because of this background, due to a weaker health capital. Their parents are also less capable of coping with adverse experiences when they occur, due to a lack of knowledge or other resources.

---

[4] Silventoinen et al. (2003).

[5] Lee et al. (2018).

Bifurcation Points and Symmetry Breaking Exist in Life Cycle Development

Although bifurcation and symmetry breaking are more common in a social context, examples can also be found from during the individual human life course. A typical example is the reception and processing of incoming information, during which an individual has to weigh the incoming information against previously received information and current images and attitudes vis-à-vis the surrounding world. Such situations are sometimes ambiguous, in the sense that more than one picture of the environment may be consistent and compatible with the incoming information.[6] It is consequently possible to imagine two almost identical individuals who receive the same incoming information but nonetheless end up with quite different pictures of the world around them.

At a higher level of perception—mindsets—young people's attitudes to education are to some extent governed by the degree of success that they experience in school work. Some less successful pupils at some stage develop an anti-education culture, which of course worsens their performance further. It is more common among boys than among girls, being part of a masculinity ideal, but there are other background determinants.[7] This represents a bifurcation point, where the less fortunate choice increases the risk of school failure, with associated elevated risks of psychic ill-health or criminal behaviour.[8]

If mechanisms generating inequality were limited to these individually based ones, overall inequality would be constrained by the human constitution—physiology, errors in perception, irrationality, and so on. It is in the social arenas where human beings interact that these naturally given bounds can be transcended and stark inequalities can develop. Turning therefore to the genesis of inequality at the social level, a number of mechanisms can be identified.

Symmetry Breaking is Pervasive in Social Systems

Different forms of symmetry breaking have been identified in previous chapters:

- In bilateral bargaining relationships, which comprises a wide variety of exchange relations in different societies, an egalitarian equilibrium in which both parties have equal power is inherently unstable. This instability is fundamental, given the ubiquity of bilateral social relations.
- Herd behaviour, in which individuals weigh their own perceptions against signals received from other members of the collective, can lead to different outcomes in identical populations, depending on which signals happen to be sent at the beginning of the decision-making process (path dependence).
- Social classes can develop on the basis of visible traits, even when these traits are irrelevant to performance in the social arena.
- Wealth concentration develops in a market economy even when starting from an equilibrium with identical individuals and identical assets.

---

[6]Molander (1986).

[7]See, for example, Warrington et al. (2000) (UK data).

[8]Gustafsson et al. (2010).

- In a market economy, two essentially identical regions may develop a strongly asymmetric relationship when transport costs decrease and pass a critical value. Similar asymmetric relationships can emerge at the global level under the influence of falling trading costs, human capital investments or the globalisation of financial markets.

The common denominator of these mechanisms is the dynamic instability of equilibria characterised by equality of resources and power among the parties involved.

Individual Inequalities Are Magnified by Positive Feedback Mechanisms in Social Systems

In real life, individuals are not identical, nor do regions have the same natural prerequisites for social and economic development. But the dynamic instability of the equilibria exemplified above has the consequence that inequalities in outcomes tend to be completely out of proportion with differences in performance underlying or justifying these inequalities.

This statement can be made more precise by studying the statistical distributions pertaining to outcomes at different stages of the transition from natural towards social variables:

- Natural (innate) differences between human beings are typically normally distributed, or possibly lognormally with a limited skewness. This is well illustrated by the body height example above.
- School performance exhibits a more skewed distribution with a constant or slightly increasing, but still moderate, spread (Sect. 4.2).
- The spread of labour income increases during working life (Sections 3.4, 4.4).
- The distribution of earnings is more skewed than the distribution of consumption (Sect. 6.5). Consumption is closer to the individual and subject to the limitations that this fact implies, even if luxurious habits may lead to high consumption levels.
- The distribution of wealth, even more distant from natural endowments, exhibits stark skewness.

In summary, inequality is to a considerable extent a social artefact. A succinct wording at a high level of abstraction would be: *The more social, the more unequal.* When observed differences are explained or justified with reference to differences in ability and effort without taking into account the dynamics of individual capital (human, etc.) and the dynamic instability of social interaction processes, the effect of ability and effort will necessarily be exaggerated.

Combining Categories of Mechanisms

In Chap. 1, mechanisms that generate inequality were sorted into the categories of *externally* generated, *internally* generated and *mixed*. In Table 9.1, this categorisation is intersected with the substantial categories *individual*, *social* and *spatial*. Some mechanisms appear in more than one slot, depending on the

Table 9.1 Summary of sources of inequality by origin and category

Category Source	Individual	Social	Spatial
External	Inheritance: material, financial, social, genetic (Sect. 3.1)	Segregation (Sect. 6.3) Discrimination (Sects. 5.2, 6.3)	Natural resources (Sects. 7.1, 8.1) Political decisions (Sects. 7.1, 8.1) Existing urban and regional structure
Mixed	Human capital formation (Sects. 3.2, 3.3, 3.4, 4.2) Health capital formation (Sects. 3.2, 3.3, 3.4, 4.3) Epigenetic processes (Sect. 3.1)	Segregation (Sect. 6.3) Discrimination (Sects. 5.2, 6.3) Conditioned choices (Sect. 6.3)	Localisation of investments (Sects. 7.1, 8.2) Labour markets (Sects. 7.1, 8.2)
Internal (symmetry breaking)	Personality development (Sect. 3.3) School failures (Sect. 3.3)	Negotiation (Sects. 5.1, 6.1) Market-generated inequality (Sects. 5.6, 6.5) Networks (Sects. 5.4, 6.4) Herd behaviour (Sects. 5.3, 6.3)	New cities (Sects. 8.1, 8.2) International division of labour, trade patterns (Sects. 7.2, 8.2)

perspective. Social and urban structures are inherited, and in this sense external, but viewed during their process of formation or development, they are internal or mixed.

9.2 Ethics

The stylised facts listed above show that a significant part of the current distribution of income, wealth and power is the result of chance and positive feedback mechanisms. There is a tendency for inequality to grow as a matter of course. This can be assumed to have consequences both for ethics—what is perceived as equitable, fair or just—and for public policymaking.

One possible conclusion from the cited research on inequality would be an attitude of inaction. If inequality has a tendency to grow autonomously, perhaps the appropriate reaction is to acquiesce and let nature follow its own course. But the patterns observed are regularities, not natural laws. As highlighted in the previous section, the dominant part of inequalities that we observe in a social context has a social origin, and the further away from a truly natural basis a certain mechanism is located, the higher the level of inequality tends to be. Societies with very similar material bases and levels of development, such as the OECD countries, exhibit highly different levels of inequality. There is ample room for ethical and political choice in these matters.

Further, the consequences of differences in natural givens, both at the individual and at the collective level, can be offset or sometimes even eliminated.

Human Societies

There exists a large, sophisticated literature on individual and collective choice, dealing with fundamental human problems of cooperation, conflict and negotiation. Starting from various assumptions, sometimes called *axioms*, researchers have proved results on the existence or non-existence of solutions to such decision problems. A striking property of these models is their level of abstraction: only seldom is there any explicit mention of the fact that they refer to human beings. A common observation from the empirical literature is that participants deviate from behavioural patterns predicted by the models—they cooperate more than expected, they do not exploit other players to the extent that they could, and so on. In short, there seems to be a moral dimension to human decision-making not captured by mainstream models.

The species *Homo sapiens* has evolved over a period of around six million years. Until recently, life was spent in small groups struggling for survival under relatively harsh conditions. On the basis of extensive studies of social behaviour among our closest primate relatives and among children, Michael Tomasello, in *A Natural History of Human Morality*, has reconstructed how humans developed moral thinking from simple cooperative behaviour.[9] A crucial step was the development of joint intentionality, a mutual understanding of the norms governing the partners to cooperation. A second step involved the development of collective intentionality, which became necessary as the progressive division of labour created more complex societies. In a later book, *Becoming Human*, Tomasello traces these stages in the development of children—joint intentionality occurring around the age of nine months, collective intentionality at the age of three, and a sense of responsibility and autonomy at the age of six to seven years in the normal child.[10]

If one accepts this picture of human morality, the diagnosis of what ails a society characterised by high inequality is that its design goes against the grain of human nature. A concrete analogy would be the basic human drive to search for calories—food rich in energy, such as fat or sweet fruits. This drive developed under conditions of more or less permanent undernourishment and served an evolutionary purpose, but in a society with a surplus of easily available calories, it increases the risk of overweight and obesity and becomes dysfunctional. Economic and cultural evolution clashes with human nature. And the difference is, of course, that human evolution is a given, whereas the results of social and cultural evolution are open for discussion, and can be changed.

It is of some interest that steps have been taken to integrate moral thinking into game-theoretic models, with the aim of investigating how morality can develop in an originally amoral environment. It turns out that preferences that combine self-

---

[9]Tomasello (2016).

[10]Tomasello (2019).

interest with morality of a "Kantian" flavour are evolutionarily stable under certain conditions, whereas preferences resulting in other types of behaviour are not.[11]

Philosophical Approaches: Legitimacy

Political philosophy has approached the problem of inequality from different angles. One tradition revolves around the concept of *legitimacy*. The common aim here is to determine what differences in resources and power in a society can be perceived as legitimate by the population at large.

Under archaic regimes, which dominated human societies until the modern era, the social order is assumed to be of divine origin and there is no need for justification by arguments. Following the Enlightenment, it became necessary to justify the position and wealth of a ruling elite by referring to excellence in moral or other respects. The concept of *desert* becomes central to this discussion. In industrialised countries, this question is strongly linked to performance in the labour market and, more generally, to the economic sphere. For instance, a survey of attitudes to legitimate differences in labour income from nine countries shows broad agreement on higher pay for high-status, elite occupations, but different opinions on how large differences are acceptable.[12] Predictably, older persons with a higher socioeconomic status and conservative preferences preferred larger differences (but not lower pay for ordinary jobs).

In *Giving Desert its Due*, Wojciech Sadurski has analysed the concept of *desert* with the aim of providing a reasonably concrete and operative delineation.[13] Sadurski ties desert to *responsibility*: basically, a person can claim to deserve only those resources that were acquired through acts for which he or she can be held responsible. This echoes a long liberal tradition going back to John Locke, who claimed that it was work that conferred on the individual the right to property.[14]

The consequences of Sadurski's definition, although footed in a liberal tradition, are rather radical. Against the backdrop of the research that has been presented in previous chapters and summarised in Table 9.1, only a limited fraction of total income or assets in fact satisfy the basic condition of responsibility, the criterion of legitimacy. Inherited property is ruled out, as are large parts of other fruits of inheritance from the family background. Windfall profits from chance events similarly lack legitimacy, as do the results of symmetry breaking and positive feedback mechanisms. Symmetry breaking implies that individuals who are identical with respect to skills and diligence and who start from equal conditions can end up with vastly different outcomes, as a result of the dynamic instability of their interaction. In reality, individuals are not identical, but the presence of dynamic instability will cause differences in outcomes to be completely out of proportion with differences in talent, effort or personal characteristics of relevance to the result.

---

[11] Alger and Weibull (2016).

[12] Kelley and Evans (1993).

[13] Sadurski (1985), in particular Chap. 5.

[14] Locke (1690), §§ 27, 40.

The Role of Genes

Because of the role that genes traditionally have played in debates on nature versus nurture and on the room for manoeuvre open to policy interventions, it seems appropriate to present an estimate of the importance of the genetic factor, according to the current status of knowledge. There are two main approaches to this question, one "from above" and one "from below". The former approach relies on population data on families—parents, children, siblings, twins and adopted children—and computes intergenerational elasticities, sibling and twin correlations, etcetera, for outcome variables such as cognitive skills or incomes. The second, more recent, approach exploits information on the human genome, identifies candidate genes for the outcome variables and computes the share of the observed variation that can be assigned to these genes.

Both approaches have their strengths and weaknesses. In a broad survey of the traditional, family-based approach, Björklund and Jäntti compare four different approaches—intergenerational mobility analysis, intergenerational intervention effects, sibling correlations, and the equal-opportunity approach.[15] The first describes how income and education in parents and offspring are correlated. The second estimates the causal effect of an intervention concerning parental income or education on their children's corresponding outcome. The sibling-correlation approach estimates the share of total inequality that can be attributed to factors shared by siblings. This represents the combined effect of genes and environment shared by siblings. The equality-of-opportunity approach tries to identify general factors that are important for children's outcomes and that children cannot be held responsible for. These approaches tend to yield quite different estimates. For instance, the share assigned to family background in sibling studies is substantially higher than that based on intergenerational mobility estimates.

Similarly, the results from genome-based studies must be considered carefully. An outcome as complex as cognitive skills depends on a large number of genes that interact, and there is some uncertainty as to whether the full ensemble of candidates have been identified. Traditionally, this method has been hampered by cost restrictions and the resulting small populations, but rapid methodological improvements in recent years have reduced this problem. The previously cited study by Lee et al. comprises 1.1 million individuals.

With the necessary caveats, it is interesting to compare results from these radically different approaches. The human genome knows no borders, so an analysis from above should include what the country of origin explains. The country of origin, including its income distribution (over which the individual has no influence), explains more than a half of the total variation.[16] Within the country, a standard model is

$$Y = gG + sS + uU,$$

[15] Björklund and Jäntti (2020).
[16] Milanovic (2015).

where Y is the outcome variable, G represents the genetic factor, S the environment shared by siblings and U the non-shared environment. Björklund et al. report $g^2 = 0.2$ and $s^2 = 0.16$ for boys, and $g^2 = 0.13$ and $s^2 = 0.18$ for girls.[17] These estimates consequently leave at most $0.2 \times 0.5 = 0.1$ for the genetic factor in a global perspective.

Lee et al. performed a large-scale genetic association analysis of educational attainment in a sample of approximately 1.1 million individuals and identified 1271 independent genome-wide-significant single-nucleotide polymorphisms. A joint analysis of educational attainment and three related cognitive phenotypes generated polygenic scores explaining 11 to 13 per cent of the variance in educational attainment and 7 to 10 per cent of the variance in cognitive performance.

Although these results should be cited with the necessary provisos, it is suggestive that the figures are of the same order of magnitude, having radically different origins. The genetic factor seems to account for around a tenth of cognitive skills and related outcomes such as educational attainment or incomes.

What does this imply for ethics and the policy discussion? Not much, actually. Even if the effects of the genetic factor are limited, and those related to social factors are correspondingly large, not all social factors are amenable to policy interventions. Further, the residual includes the chance factor, which can be limited but not eliminated. On the other hand, the fact that a certain outcome is genetically "determined" does not mean that policy interventions are excluded. As pointed out by Goldberger, many visual impairments have a genetic origin but can easily be corrected using eyeglasses.[18] The important criterion for deciding on an intervention is not whether a problem identified is of a genetic origin or not but whether it can be accommodated with a reasonable efficiency-to-cost ratio.

In conclusion, the role of genetic factors seems to have been overrated, both substantively and when it comes to policy relevance.

Philosophical Approaches: Contract Theory
Contract theory has a long tradition in political philosophy, going back to classical Greek philosophers and to Roman Lucretius, continuing through the Middle Ages and Hobbes in the seventeenth century, to be revived, perhaps somewhat unexpectedly, in the 1970s by such diverse thinkers as John Rawls and Robert Nozick. The basic approach of contract theory is intellectually appealing. Everyone is stripped of his or her identity and status, and the political problem is reduced to the question, "What kind of society would a population in such an anonymous situation prefer?"

The idea that participants in a discussion about basic principles must disregard their own situation is reminiscent of Adam Smith's impartial spectator, but the latter concept is wider.[19]

---

[17] Björklund et al. (2005).
[18] Goldberger (1979).
[19] Sen (2009).

The outcome of the constitutional discussion on the social contract will of course depend on what assumptions are made about both the participants and the environment. Thomas Hobbes, writing *Leviathan* during a civil war, was primarily interested in how to guarantee basic security to life and limb and was pessimistic about the ability of citizens to resolve their differences in peace. Jean-Jacques Rousseau, writing *The Social Contract* and *Discourse on Inequality* a century later in a French nation as yet untouched by revolution, took basic peace for granted and spent more time on the problem of inequality.

John Rawls and Robert Nozick, two of the main contributors to the philosophical debate of the 1970s, both labelled themselves "liberals". Rawls uses a contract approach in *A Theory of Justice*, while Nozick in *Anarchy, State, and Utopia* mentions contracts only in passing and sets out to design the perfect constitution. The two authors differ less in their approach than in their solutions, however. Nozick's solution is process-oriented; he attempts to lay down principles that are permanent, in the sense that the results of any actions taken in accordance with these principles will be legitimate. Rawls includes outcomes in the criteria by which a society should be judged[20]:

> Social and economic inequalities are to be arranged so that they are both (a) to the greatest benefit of the least advantaged and (b) attached to offices and positions open to all under conditions of fair equality of opportunity.

Nozick's solution is open-ended, whereas Rawls introduces feedback from the outcome by requiring measures to be taken in order to ensure that the least advantaged are better off than following any other policy. Against the backdrop of the pervasive instability of social processes that has been illustrated in preceding chapters, it can be concluded that Rawls' solution can possibly work, whereas Nozick's is destined to fail. Any feasible solution of the policy problem has to satisfy some basic stability requirement, and it is impossible to stabilise an unstable system without recourse to some sort of feedback rule. Rawls' solution does satisfy this design requirement, which is not to say that it is necessarily the only or even the best solution; there may be other reasons for criticising it.

It seems that contract theory has come to stay in popular philosophy, and its focus is relevant to issues of distribution and inequality. This approach is intellectually appealing, but there is a major design fault: we do not live in a contract situation. Even if it is possible to lead a discussion under the abstract assumptions of a contractarian situation, these assumptions will be abandoned as soon as the discussion becomes operative and approaches current problems regarding social and economic policy. It is unlikely that consensus will be reached.[21] A more realistic approach is to scrutinise the arguments for and against inequality at a more concrete level, on a hopefully common ground of established facts.

---

[20]Rawls (1971), § 11.

[21]For a developed critique of the contractarian approach, see Sadurski (1985).

9.3 Policy Arguments

The traditional view of inequality in economics is that it is beneficial. Inequality, very broadly speaking, creates incentives, and incentives stimulate effort, which benefits growth. Political ambitions to even out inequalities may be well intentioned, but are harmful and may even be counter-productive, benefiting better-off households. A widespread metaphor has been *the leaky bucket*, conveying the image that transferring resources to low-income households is like carrying water in a leaky bucket; resources are squandered in the process, and decisionmakers must ask themselves whether the operation is worth the cost.[22] The literature on connections between inequality and different social outcomes has expanded beyond this simple metaphor in recent decades, and this section provides a survey of main results.

Inequality and Economic Growth
The main arguments for why inequality should benefit economic growth are based on links with *savings*, with *investments in human capital*, and with *entrepreneurship*. Given that high-income earners save more in relation to their income, the total savings will be higher if income distribution is more skewed, which should lead to higher investments and a higher growth rate. Investments in human capital have a higher payoff if salaries in professional occupations are higher than average wages, which would encourage more young people to invest in education. Finally, it is easier to become a self-employed entrepreneur if income is high enough to permit some extra saving before starting, and high incomes for entrepreneurs, whether market-based or subsidised, create incentives in the same way as high salaries to professionals.

So, what do the data say? A survey by Neves & Silva covers 30-plus articles on the connection between inequality and growth, to which can be added analyses at the IMF, the OECD and the CESifo.[23] A quick look at the literature gives a mixed impression. Some studies find a positive relationship between inequality and growth rate, others find a negative one, and others still fail to establish any statistically significant link. The sample of countries varies: some studies are based only on industrial countries, while others also include developing countries and emerging economies. The time period studied is in most cases 1960–2000. The measure of inequality varies; most studies are based on the Gini coefficient, whereas a smaller group use both the Gini coefficient and more disaggregated measures, such as the income shares of quintile or decile groups.

The results can be summarised in the following way. In the large group of studies based on the Gini coefficient, some researchers do not find any significant result at all, while others find a positive effect of inequality on growth in rich countries and a negative one in poor countries, and others still find a negative relationship. In the smaller group, where the analysis is based on disaggregate measures of inequality,

---

[22] Okun (1975). A counter-metaphor was launched by Korpi (1985).

[23] Neves and Silva (2014), Dabla-Norris et al. (2015), Cingano (2014) and Fuest et al. (2018), respectively.

all studies with one exception find that inequality harms growth; in the deviating study, based on 21 developed countries, inequality is beneficial at the top and negative at the lower end.[24]

How should these results be interpreted? As pointed out in Chap. 2,[25] the Gini coefficient is a blunt instrument for measuring inequality; countries with very different income distributions, and therefore very different incentive structures, can have the same Gini coefficient. It is therefore no surprise that studies based on this index do not produce robust results. More confidence must be placed in the more detailed studies.

By way of example, the table below shows the results from the analysis by Dabla-Norris et al., based on a large sample of both developing and industrialised countries (Table 9.2).

The statistical analysis shows that higher income shares for the lower quintile groups *increase* the economic growth rate. A higher income share for the highest income quintile group *decreases* the growth rate. The results from Cingano (2014) are similar.

How should these results be interpreted against the backdrop of the arguments presented at the beginning of the section? There are several aspects. In the first place, human capital accumulation has replaced physical capital accumulation as the prime engine of growth.[26] At the beginning of the industrial era, physical capital accumulation was a prime source of growth, and inequality facilitated the channelling of financial resources into this process of accumulation given the higher savings rate in high-income strata. As human capital has gradually taken over the role of the prime mover, the adverse effects of inequality have gradually become dominant.

The importance of human capital and its distribution has been confirmed in a study by Castelló and Doménech.[27] As shown above, the negative effect of income inequality measured by Gini coefficients on economic growth rates is not robust, but there is a robust negative effect of human capital inequality (measured by the Gini coefficient) on economic growth rates. In short, higher educational inequality is associated with lower investment rates and lower income growth.

As regards the negative correlation between the income share of the highest quintile group and growth, it must be realised that the cited argument about savings rates is valid only in a closed economy. Given a globalised financial market, there is no guarantee that the resources saved will stay in the country, enhancing economic growth. Even if they do stay in the country, the developed growth model by Stiglitz that makes the distinction between wealth and capital can explain a negative relationship.[28] If a higher share of total wealth is concentrated in the upper strata,

---

[24]These studies are Persson and Tabellini (1994), Clarke (1995), Perotti (1996), Voitchovsky (2005), Castelló (2010), Cingano (2014) and Dabla-Norris et al. (2015).

[25]See Fig. 2.2.

[26]Galor and Moav (2004), Galor (2011).

[27]Castelló and Doménech (2002).

[28]Stiglitz (2016). See Sect. 6.5.

Table 9.2 Regression of GDP growth rate on income distribution and other background variables. Source: Dabla-Norris et al. (2015)

Variables	Dependent variable: GDP growth					
	(1)	(2)	(3)	(4)	(5)	(6)
Lagged GDP growth	0.145*** (0.033)	0.112*** (0.030)	0.118*** (0.031)	0.113*** (0.031)	0.097*** (0.030)	0.114*** (0.031)
GDP per capita level (in logs)	−1.440*** (0.361)	−2.198*** (0.302)	−2.247*** (0.307)	−2.223*** (0.308)	−2.122*** (0.304)	−2.222*** (0.307)
Net Gini	−0.0666* (0.034)					
1st quintile group		0.381** (0.165)				
2nd quintile group			0.325** (0.146)			
3rd quintile group				0.266* (0.152)		
4th quintile group					0.0596 (0.180)	
5th quintile group						−0.0837* (0.044)
Constant	17.34*** (3.225)	18.82*** (2.579)	18.12*** (2.713)	17.45*** (3.058)	19.41*** (4.203)	25.32*** (3.496)
Country fixed effects	Yes	Yes	Yes	Yes	Yes	Yes
Time dummies	Yes	Yes	Yes	Yes	Yes	Yes
# observations	733	455	455	455	455	455
# countries	159	156	156	156	156	156

a larger fraction will also be channelled into non-productive investments—"land", in the generalised sense.

As for the *private incentives* to invest in human capital, it is true that there are large variations even within the relatively homogeneous group of OECD countries. For instance, the Scandinavian countries have lower wage premia coupled to tertiary education than the OECD average, but the fraction of adults with tertiary education is nonetheless higher than the OECD average in these countries. Obviously, the choice of educational level is determined by other factors besides expected future wages. Subsidised higher education is one way of improving the trade-off in the presence of low wage gaps, and general attitudes to inequality are important.[29] Further, the analyses of the OECD are incomplete in the sense that they draw the line at the age of 65. Higher wages normally imply higher pensions, and tertiary education is linked to higher expected longevity, so the higher pension is also received over a longer period.

The third argument, related to *entrepreneurship*, is somewhat complex to disentangle. Aghion et al. (2019) establish a correlation between innovations and top-income inequality, but draw the conclusion that causality goes from innovation to inequality. More generally, the idea of single individuals saving for the possibility of becoming self-employed in a company that eventually grows into a large corporation is not representative of industrial development. Many factors determine the business climate in a country—the general level of education in the workforce, the quality of the infrastructure, absence of corruption and others—and income and wealth inequality would be a less relevant factor in this context.[30]

Inequality and Social Mobility

There is wide consensus across the political spectrum that intergenerational mobility is positive. The idea that all children, as far as possible, should have equal chances to form their own destiny is fundamental to liberal society. There is also strong evidence of a correlation between inequality and low mobility,[31] even if not all statements about mobility are robust with respect to the choice of mobility measure.[32]

As illustrated in Fig. 3.2, the system of education, and more generally the infrastructure for investment in human capital, is a decisive factor in this context. The attitude of the economically dominant elite towards investments in human capital affects not only the distribution of capital in society but also the character and rate of economic growth. In agriculturally dominated societies, landowners tend to be more negative to raising the general level of education than industrial capitalists, because knowledge is a less complementary factor of production in agriculture

---

[29] Björklund and Freeman (2010).

[30] For examples of important factors, see, for example, World Bank (2019). A broad critical discussion of political arguments used in small business policy around the world can be found in Atkinson and Lind (2018).

[31] See Sect. 3.1.

[32] Jäntti and Jenkins (2015).

than in industry—a fact that contributed to the divergence of growth rates not only at the global level but also within the United States.[33] But also later, after industrialisation, the character of investments has been affected by inequality and attitudes. A high level of inequality raises the level of investment in industries that are intensive in physical capital, whereas lower levels of inequality tend to benefit industries that are intensive in human capital.[34]

Inequality and Trust

High inequality correlates with low trust in a number of studies.[35] It appears that inequality in the lower part of the distribution is particularly important for this connection.[36] Correlation is different from causation, however, and an attempt to establish a causal connection between two aggregate indicators of this kind is not meaningful. By breaking down the trust complex into components, it is possible to identify possible channels of causation. One important channel goes via the perceived mobility in a society—the correlation between the fraction of a population that perceives society to be meritocratic and the level of trust is even higher than the correlation between inequality and trust—and as we have seen, mobility is linked to inequality.[37] There is also a link between trust and corruption, which goes in both directions, however, and the stronger link appears to go from trust to corruption.[38]

Trust in turn is positively associated with a number of important welfare indicators—resilience (both individual and social),[39] health,[40] perceived wellbeing,[41] and economic growth. That trust should be positive for economic growth is an old observation, going back to Adam Smith and John Stuart Mill, and in modern times to Kenneth Arrow.[42] Concrete expressions of high trust are lower transaction costs,[43] increased possibilities of delegation,[44] and the direct effect on investments.[45] Empirical tests confirm the positive effects of trust on economic growth, but only in democracies.[46]

---

[33] Galor et al. (2009).

[34] Erman and te Kaat (2019).

[35] Uslaner and Brown (2005), Bjørnskov (2007), Jordahl (2009), Rothstein (2011), Barone and Mocetti (2016).

[36] Gustavsson and Jordahl (2008).

[37] Larsen (2013), p. 108.

[38] You (2018), Robbins (2012).

[39] Helliwell et al. (2018).

[40] Kawachi (2018).

[41] Goff et al. (2018), Helliwell et al. (2018).

[42] Arrow (1972).

[43] Williamson (2000), Bjørnskov and Méon (2015).

[44] Bloom and van Reenen (2010), Gur and Bjørnskov (2017).

[45] Zak and Knack (2001).

[46] Dincer and Uslaner (2010) (US data), Bjørnskov (2018).

Inequality and Health

The possible connections between inequality and health have been a contentious issue in recent decades. There is consensus that there is a positive connection between income and health, both at the individual and at the aggregate level. Health improves with income, but there is also a saturation effect.[47] The question is whether growing inequality, for a given average level of income, has negative effects on average health. If inequality is measured by income spread, as the coefficient of variation (standard deviation divided by mean), there is a purely mechanical effect of increasing inequality on average health.[48] What has been debated is whether there is also a substantive effect from inequality on health. Research groups have come to varying conclusions.[49]

Given what is known about the effect of ranking on health in other primate species, it is likely that factors in the social environment such as status, both in early life and adulthood, affect morbidity and mortality risk in humans. Studies of other social mammals indicate similarities across species.[50] The fact that perceived economic status is a better predictor of health than objective measures of economic standard supports this line of reasoning.[51]

The difficulty of separating what was called the mechanical effect above from the genuine status or inequality effect makes the identification a challenge. The combined effect is doubtless stronger in the lower strata of the population, and the mechanical effect probably dominates the picture.

Inequality and Conflict

The nexus between inequality and conflict is another relevant policy aspect. The term *conflict* covers a wide spectrum of social phenomena. When Persson and Tabellini register a negative correlation between inequality and economic growth, they interpret this as the effect of vociferous requirements for transfers that may be costly.[52] As explained above, there are stronger and more direct links from inequality to low growth rates, but the mechanism alluded to by the authors may of course still be active.

Another link of relevance to the discussion is that between inequality and crime. Focus in the debate has been on violent crime—robbery, murder—while economic crime may of course also be affected. A major issue in the social science literature on this topic is the question of whether the driving force behind crime is low standards of living or high inequality, that is, whether the propensity to commit crime is affected by absolute or relative standards. In a broad survey based on data and

---

[47] In mathematical terms, the first derivative is positive, the second is negative.

[48] This follows from a straightforward differentiation with respect to the standard deviation σ.

[49] Macinko et al. (2003), Lynch et al. (2004), Subramanian and Kawachi (2004), Kondo et al. (2009), Wilkinson and Pickett (2009), Fritzell et al. (2013), Pickett and Wilkinson (2015), Bergh et al. (2016).

[50] Abbott et al. (2003), Sapolsky (2004), Snyder-Mackler et al. (2020).

[51] Adler and Ostrove (1999), Singh-Manoux et al. (2003).

[52] Persson and Tabellini (1994).

research from the United States, Daly shows that the relative perspective rather than the absolute provides the best explanation.[53]

Turning to conflicts of higher dimensions, the previously cited study by Milanovic on the correlation between the inequality extraction ratio and the risk of violent conflict points in the same direction.[54] The fact that the inequality extraction ratio and not the Gini coefficient is the best predictor of conflict underlines the relative nature of the problem at a deeper level; it is not inequality in itself but inequality in relation to what the current material situation permits that determines the legitimacy of the regime.

Regional Aspects

It has been shown in Chaps. 7 and 8 that urban and regional development exhibits parallels with individual development and that similar mechanisms generating inequality can be found in the two areas. Inherited natural conditions, chance and political decisions in a distant past are factors that confer a certain degree of arbitrariness on existing regional structures. The norm that living conditions and opportunities for children should as far as possible be independent of which family they have been born into can be extended to neighbourhoods and regions. Arguments for redistribution can consequently be formulated and tested also in the regional context.

Public power is necessarily exercised across more than one level of government—how many depends on the size of the country, historical tradition and other factors. The standard division of labour within the public sector is to assign responsibilities to the level where public goods are naturally produced.[55] Local infrastructure such as streets or water and sewerage are municipal responsibilities, while defence and telecommunications are necessarily national concerns. This simple principle gives rise to a number of questions, though.

Not all public goods are so easily categorised along the spatial dimension. Education is an important example, where goals and restrictions are formulated at the national level but production is necessarily local. The human capital produced in schools represents a national collective asset. The fact that children move when they grow up—in Sweden, more than half of the adult population lives outside the municipality in which they grew up[56]—creates a collective-action problem. A municipality that invests generously in schooling will see the fruits of more than half of its investments disappear.

Further, the national level in many areas sets a minimal allowable standard for public services. This imposes costs on the local level, and the implementation of nationwide systems should not be dependent on the local ability to pay. Some form of fiscal transfers is therefore necessary. This is not a case of redistribution, properly

---

[53] Daly (2016). The author focuses on murder as an indicator, as it is a well-defined crime for which the room for statistical judgement is limited.

[54] Milanovic (2013).

[55] Boadway and Mörk (2004).

[56] Data from Statistics Sweden.

speaking; rather, it should be seen as a corrective to a government failure that would otherwise occur.

The level of ambition and the detailed design differ between countries, depending, for instance, on the size of public commitments and the manner in which rights of taxation are distributed.[57] Because units at the local and regional level differ with respect to both costs and revenues, some form of equalisation on both sides of the budget is necessary. Such a system of equalisation needs to take into consideration the distribution of natural resources, demography, urban structure and other factors that the local level cannot affect in the short run, but no others, in order to be incentive-compatible.[58]

The central level affects living conditions in the regions in other ways than via intergovernmental transfers. The quality of national infrastructure networks across the territory is a case in point. Given the importance of human capital formation to social development and the spontaneous drift of the younger generation towards educational centres and metropolitan areas, access to higher education is another potential lever for equalising opportunities. There are indications that the proximity to institutions at the tertiary level does affect recruitment to higher education.[59]

In the same way as policies of redistribution in general have been scrutinised with respect to incentives, the question has been raised as to whether policies aimed at equalisation of living conditions between regions may be harmful to economic growth. Is spatial agglomeration in fact positive? The answer turns out to be highly dependent on the level at which agglomeration is measured, and what measure is used.[60] In general, the answer for a given region will vary over time, depending on the stage of industrial development.[61]

9.4 Policy Design

This section collects some basic design principles that follow from the previous analysis.

Feedback Is Necessary

Both theoretical and empirical evidence shows that inequality is inherent in human affairs, stemming from the dynamic instability of egalitarian distributions in a range of recurrent situations. Such instability cannot be controlled by any apriori rule that does not take the outcome into account. Feedback from the outcome is a prerequisite for successful containment of inequality within socially acceptable boundaries.

---

[57] For international overviews, see Boadway and Shah (2007) and Boadway and Eyraud (2018).

[58] For an example, see Tingvall (2007).

[59] Drucker and Goldstein (2007), Drucker (2016) (U.S.), Eliasson (2006) (Sweden).

[60] Gardiner et al. (2010).

[61] Potter and Watts (2010).

Feasibility of Equalising Policies

If the consequences for inequality of some factor are considerably negative, the character of this factor—environmental, genetic—is not of primary importance. What is important is how the negative effects can be mitigated or eliminated, and at what cost in the wide sense—if there is a cost. The calculation of the ratio of benefits to costs needs to consider both immediate and long-term effects, as well as side effects in other areas than the one directly concerned.

Which Inequality Is Important?

A political ambition to contain or reduce inequality must be specified in order to become operative. Inequality refers to a distribution, and policy design requires that the focus of equalising policies be made precise. As illustrated in Chap. 2,[62] different profiles may give rise to the same Gini coefficient, so keeping this coefficient within bounds is insufficient as a political target.

Repeatedly, the importance of reducing inequality among the lower strata has emerged in the analysis. Educational policy is important to social mobility, and it is the lower strata that are sensitive to thresholds in the system of education.[63] Generalised social trust is dependent on inequality, but this dependence is particularly strong among lower strata.[64] The level of trust is an important component of subjective wellbeing, and again, the lower income strata, or more generally vulnerable households, come out as the decisive group behind the general level of subjective well-being and, a fortiori, of political legitimacy in a society.[65]

In summary, the position of the lower strata as regards living standards and opportunities is decisive. This must be kept in mind in the design of redistribution policies. A diffuse goal of containing or reducing inequality in a general sense is inadequate.

Front-Loaded Interventions

A recurrent theme in the description and analysis of lifecycles is the importance of early stages in individual development. Public interventions should therefore be frontloaded—whether one considers advice in maternal care, preschools, schools or other policy instruments pertaining to the early-life environment. Early interventions have stronger effects and are active over a longer period, so cost-benefit analyses systematically yield a higher payoff for this category. The problem here is the targeting of the interventions: policies must be selective, and groups in focus must be identified.

Social Traps

Traps born out of behavioural rules that are contingent on the decisions of others require coordinated interventions in several domains—education, housing, income

---

[62] Fig. 2.2.

[63] Fig. 3.1.

[64] Gustavsson and Jordahl (2008).

[65] Esaiasson et al. (2020).

security and possibly other policy areas. This requires coordination between different parts of the social service administration, which can be difficult to accomplish when strict data integrity rules are to be respected.

Human Versus Physical Capital

The importance of human capital is increasing, not only for growth and development in general but also for reducing rising inequalities. As noted above, in countries with a higher level of equality, investments in human-capital-intensive industries are relatively more important,[66] so causality between human capital and equality goes in both directions. It must also be kept in mind that human capital is to a considerable extent a collective asset,[67] so there is a latent risk of underinvestment in cases where the public sector does not take full responsibility for this component of national wealth.

The surge in investments in artificial intelligence represents a potential threat to the equalising force of human capital investments. If human skills, also at high levels, are codified and computerised, the borderline between physical and human capital is to some extent eliminated. This may imply a step backwards into the early phases of industrialisation, when key investment decisions relied on the concentration of financial capital into relatively few hands.

Spatial Heterogeneity

The case for equalisation of regional disparities is as strong as it is for the equalisation of other differences beyond the reach of individual agents in the social system. Techniques for implementing such equalisation without creating serious disincentives at the local or regional level exist.

Bibliography

Abbott, D. H., et al. (2003). Are subordinates always stressed? A comparative analysis of rank differences in cortisol levels among primates. *Hormones and Behavior, 43*, 67–82.

Adler, N., & Ostrove, J. (1999). SES and health: What we know and what we don't. *Annals of the New York Academy of Sciences, 896*, 3–15.

Aghion, P., et al. (2019). Innovation and top income inequality. *The Review of Economic Studies, 86*(1), 1–45.

Alger, I., & Weibull, J. (2016). Evolution and Kantian morality. *Games and Economic Behavior, 98*, 56–67.

Arrow, K. J. (1972). Gifts and exchanges. *Philosophy and Public Affairs, 1*(4), 343–362.

Atkinson, R. D., & Lind, M. (2018). *Big is beautiful. Breaking the myth of small business.* MIT Press.

Barone, G., & Mocetti, S. (2016). Inequality and trust: New evidence from panel data. *Economic Inquiry, 54*(2), 794–809.

---

[66] Erman and te Kaat (2019).

[67] McMahon (2009, 2018).

Bergh, A., et al. (2016). *Sick of inequality? An introduction to the relationship between inequality and health*. Edward Elgar.

Björklund, A., & Freeman, R. B. (2010). Searching for optimal inequality/incentives. In R. B. Freeman et al. (Eds.), *Reforming the welfare state. Recovery and beyond in Sweden*. University of Chicago Press.

Björklund, A., & Jäntti, M. (2020). Intergenerational mobility, intergenerational effects, sibling correlations, and equality of opportunity: A comparison of four approaches. *Research in Social Stratification and Mobility, 70*. https://doi.org/10.1016/j.rssm.2019.100455.

Björklund, A., et al. (2005). Influences of nature and nurture on earnings variation: A report on a study of sibling types in Sweden. In S. Bowles et al. (Eds.), *Unequal chances: Family background and economic success* (pp. 145–164). Princeton University Press.

Bjørnskov, C. (2007). Determinants of general trust: A cross-country comparison. *Public Choice, 130*(1–2), 1–21.

Bjørnskov, C. (2018). Social trust and economic growth. In E. Uslaner (Ed.), *The Oxford handbook of social and political trust*. Oxford University Press.

Bjørnskov, C., & Méon, P.-G. (2015). The productivity of trust. *World Development, 70*, 317–331.

Bloom, N., & van Reenen, J. (2010). Why do management practices differ across firms and countries? *Journal of Economic Perspectives, 24*(1), 203–224.

Boadway, R., & Eyraud, L. (2018). Designing sound fiscal relations across government levels in decentralized countries. *IMF working paper* WP/18/271. Washington, DC: The International Monetary Fund.

Boadway, R., & Mörk, E. (2004). Division of powers. Ch. 2 in P. Molander (Ed.), *Fiscal federalism in unitary states*. Dordrecht: Kluwer Academic Publishers.

Boadway, R., & Shah, A. (2007). *Intergovernmental fiscal transfers. Principles and practice*. The World Bank.

Castelló, A. (2010). Inequality and growth in advanced economies: An empirical investigation. *Journal of Economic Inequality, 8*, 293–321.

Castelló, A., & Doménech, R. (2002). Human capital inequality and economic growth: Some new evidence. *The Economic Journal, 112*, C187–C200.

Cingano, F. (2014). Trends in income inequality and its impact on economic growth. *OECD social, employment and migration working papers*, No. 163, OECD Publishing.

Clarke, G. (1995). More evidence on income distribution and growth. *Journal of Development Economics, 47*, 403–427.

Dabla-Norris, E., et al. (2015). Causes and consequences of income inequality: A global perspective. *IMF Staff Discussion Note* SDN/15/13.

Daly, M. (2016). *Killing the competition. Economic inequality and homicide*. Transaction.

Dincer, O. C., & Uslaner, E. M. (2010). Trust and growth. *Public Choice, 142*, 59–67.

Drucker, J. (2016). Reconsidering the regional economic development impacts of higher education institutions in the United States. *Regional Studies, 50*(7), 1185–1202.

Drucker, J., & Goldstein, H. (2007). Assessing the regional economic development impacts of universities: A review of current approaches. *International Regional Science Review, 30*(1), 20–46.

Eliasson, K. (2006). *College choice and earnings among university graduates in Sweden*. Ph.D. diss., Department of Economics, University of Umeå.

Erman, L., & te Kaat, D. M. (2019). Inequality and growth: Industry-level evidence. *Journal of Economic Growth, 24*, 283–308.

Esaiasson, P., et al. (2020). In pursuit of happiness: Life satisfaction drives political support. *European Journal of Political Research, 59*, 25–44.

Fritzell, J., et al. (2013). Cross-temporal and cross-national poverty and mortality rates among developed countries. *Journal of Environmental and Public Health* vol. 2013, Article ID 915490. https://doi.org/10.1155/2013/915490.

Fuest, C., et al. (2018). *Why the IMF and OECD are wrong about inequality and growth*. European Network of Economic and Fiscal Policy Research/ifo.

Galor, O. (2011). Inequality, human capital formation, and the process of development. Ch. 5 in
    E. A. Hanushek et al. (Eds.), *Handbook of the economics of education*, Vol. 4, pp. 441–493.
Galor, O., & Moav, O. (2004). From physical to human capital accumulation: Inequality and the
    process of development. *Review of Economic Studies, 71*, 1001–1026.
Galor, O., et al. (2009). Inequality in landownership, the emergence of human-capital promoting
    institutions, and the great divergence. *The Review of Economic Studies, 76*(1), 143–179.
Gardiner, B., et al. (2010). Does spatial agglomeration increase national growth? Some evidence
    from Europe. *Journal of Economic Geography, 11*(6), 979–1006.
Goff, L., et al. (2018). Inequality of subjective well-being as a comprehensive measure of inequal-
    ity. *Economic Inquiry, 56*(4), 2177–2194.
Goldberger, A. S. (1979). Heritability. *Economica*, New Series, 46(184), 327–347.
Gur, N., & Bjørnskov, C. (2017). Trust and delegation: Theory and evidence. *Journal of Compar-
    ative Economics, 45*(3), 644–657.
Gustafsson, J.-E., et al. (2010). *School, learning and mental health. A systematic review*. The Royal
    Swedish Academy of Sciences.
Gustavsson, M., & Jordahl, H. (2008). Inequality and trust in Sweden: Some inequalities are more
    harmful than others. *Journal of Public Economics, 92*, 348–365.
Helliwell, J. F., et al. (2018). New evidence on trust and well-being. In E. Uslaner (Ed.), *The Oxford
    handbook of social and political trust*. Oxford University Press.
Hume, D. (1739–40). *A treatise of human nature*. Several later editions.
Jäntti, M., & Jenkins, S. P. (2015). Income mobility. Chapter 10 in A. B. Atkinson, &
    F. Bourguignon (Eds.), *Handbook of income distribution*, Vol. 2, 807–935.
Jordahl, H. (2009). Economic inequality. Ch. 19 in G. T. Svendsen, & G. L. H. Svendsen (Eds.),
    Handbook of social capital. Cheltenham: Edward Elgar.
Kawachi, I. (2018). Trust and population health. In E. Uslaner (Ed.), *The Oxford handbook of social
    and political trust*. Oxford University Press.
Kelley, J., & Evans, M. D. R. (1993). The legitimation of inequality: Occupational earnings in nine
    countries. *The American Journal of Sociology, 99*(1), 75–125.
Kondo, N., et al. (2009). Income inequality, mortality, and self-rated health: Meta-analysis of
    multilevel studies. *British Medical Journal, 339*, b4471.
Korpi, W. (1985). Economic growth and the welfare state: Leaky bucket or irrigation system?
    European Sociological Review, 1(2), 97–118.
Larsen, C. A. (2013). *The rise and fall of social cohesion*. Oxford University Press.
Lee, J. J., et al. (2018). Gene discovery and polygenic prediction from a genome-wide association
    study of educational attainment in 1.1 million individuals. *Nature Genetics, 50*, 1112–1121.
Locke, J. (1690). *Two treatises on government*. Several later editions.
Lynch, J., et al. (2004). Is inequality a determinant of population health [. . .]? *The Milbank
    Quarterly, 82*(1), 5–99.
Macinko, J. A., et al. (2003). Income inequality and health: A critical review of the literature.
    Medical Care Research and Review, 60(4), 407–452.
McMahon, W. W. (2009). *Higher learning, greater good. The private and social benefits of higher
    education*. The Johns Hopkins University Press.
McMahon, W. W. (2018). The total return to higher education: Is there underinvestment for
    economic growth and development? *The Quarterly Review of Economics and Finance, 70*,
    90–111.
Milanovic, B. (2013). The inequality possibility frontier. Extensions and new applications. *Policy
    research working paper* 6449. Washington, DC: The World Bank.
Milanovic, B. (2015). Global inequality of opportunity: How much of our income is determined by
    where we live? *The Review of Economics and Statistics, 97*(2), 452–460.
Molander, P. (1986). Induction of categories: The problem of multiple equilibria. *Journal of
    Mathematical Psychology, 30*(1), 42–54.
Neves, P., & Silva, S. M. (2014). Survey article: Inequality and growth: Uncovering the main
    conclusions from the empirics. *Journal of Development Studies, 50*(1), 1–21.

Okun, A. (1975). *Equality and efficiency: The big trade-off*. The Brookings Institution.

Perotti, R. (1996). Growth, income distribution, and democracy: What the data say. *Journal of Economic Growth, 1*, 149–187.

Persson, T., & Tabellini, G. (1994). Is inequality harmful for growth? *American Economic Review, 84*(3), 600–621.

Pickett, K. E., & Wilkinson, R. G. (2015). Income inequality and health: A causal review. *Social Science & Medicine, 128*, 316–326.

Potter, A., & Watts, H. D. (2010). Evolutionary agglomeration theory: Increasing returns, diminishing returns, and the industry life cycle. *Journal of Economic Geography, 11*, 417–455.

Rawls, J. (1971). *A theory of justice*. Harvard University Press.

Robbins, B. G. (2012). Institutional quality and generalized trust: A nonrecursive causal model. *Social Indicators Research, 107*, 235–258.

Rothstein, B. (2011). *The quality of government. Corruption, social trust and inequality in an international perspective*. University of Chicago Press.

Sadurski, W. (1985). *Giving desert its due. Social justice and legal theory*. Dordrecht: D. Reidel Publishing Company.

Sapolsky, R. M. (2004). Social status and health in humans and other animals. *Annual Review of Anthropology, 33*, 393–418.

Schurz, G. (1997). *The is-ought problem. An investigation in philosophical logic*. Kluwer.

Sen, A. (2009). Introduction. In Smith, A. (1759/2009), *The theory of moral sentiments*. New York: Penguin Group.

Silventoinen, K., et al. (2003). Heritability of adult body height: A comparative study of twin cohorts in eight countries. *Twin Research and Human Genetics, 6*(5), 399–408.

Singh-Manoux, A., et al. (2003). Subjective social status: Its determinants and its association with measures of ill-health in the Whitehall II study. *Social Science and Medicine, 56*, 1321–1333.

Snyder-Mackler, N., et al. (2020). Social determinants of health and survival in humans and other animals. *Science, 368*, eaax9553 (22 May 2020).

Staub, K., et al. (2015). From left-skewness to symmetry: How body-height distribution among Swiss conscripts has changed shape since the late 19th century. *Annals of Human Biology, 42*(3), 262–269.

Stiglitz, J. E. (2016). New theoretical perspectives on the distribution of income and wealth among individuals. Ch. 1 in K. Basu, & J. E. Stiglitz (Eds.), *Inequality and growth: Patterns and policy. Volume I: Concepts and analysis*. Berlin: Springer.

Subramanian, S. V., & Kawachi, I. (2004). Income inequality and health: What have we learned so far? *Epidemiological Reviews, 26*, 78–91.

Tingvall, L. (2007). *Local government financial equalisation in Sweden*. Mimeo. Stockholm: Ministry of Finance.

Tomasello, M. (2016). *A natural history of human morality*. Harvard University Press.

Tomasello, M. (2019). *Becoming human. A theory of ontogeny*. Belknap/Harvard University Press.

Uslaner, E., & Brown, M. (2005). Inequality, trust, and civic engagement. *American Politics Research, 33*(6), 868–894.

Voitchovsky, S. (2005). Does the profile of economic inequality matter for economic growth? Distinguishing between the effects of inequality in different parts of the income distribution. *Journal of Economic Growth, 10*, 273–296.

Warrington, W., et al. (2000). Student attitudes, image and the gender gap. *British Educational Research Journal, 26*(3), 393–407.

Wilkinson, R. G., & Pickett, K. (2009). *The spirit level: Why more equal societies almost always do better*. Allen Lane.

Williamson, O. E. (2000). The new institutional economics: Taking stock, looking ahead. *Journal of Economic Literature, XXXVIII*(3), 595–613.

World Bank. (2019). *Doing business 2019. Training for reform.* Washington, DC: The World Bank.

You, J.-S. (2018). Trust and corruption. In E. Uslaner (Ed.), *The Oxford handbook of social and political trust.* Oxford University Press.

Zak, P. J., & Knack, S. (2001). Trust and growth. *The Economic Journal, 111*, 295–321.

Chapter 10
Conclusion

The general message from the research literature on the distribution of resources in a society is that inequality is no aberration from a normal state of equality; it is to be expected. This is a qualitative statement. The level of inequality is not bound by any law of nature but can be affected by policy measures.

The arguments for containing and reducing inequality are strong. Generally supported political goals such as economic growth, social mobility, health and trust are all positively correlated with equality, and the links are in important respects causal.

10.1 Economics, Politics and Institutions

When the foundations of modern social science were laid by thinkers such as Adam Smith, David Ricardo and Auguste Comte, the integration of a multitude of perspectives, including the political, was taken for granted. Ricardo named the discipline *political economy* for a reason. For a long time, economics lost this perspective under the dominance of the general-equilibrium paradigm. In recent decades, some balance has been restored by the expansion of the institutional approach in the intersection of political science and economics. Typical questions in focus are the evolution and transformation of institutions, as well as the effect of institutions on economic performance.[1]

It is a step forward that the importance of political variables has been recognised. Nonetheless, there is a risk that large-scale statistical analyses of links between institutional characteristics and economic growth or general well-being yield misleading results. It is difficult to reduce complex societal processes to a limited set of

---

[1] For some typical examples, see North (1990), Persson and Tabellini (2003) and Acemoglu and Robinson (2006).

© The Author(s), under exclusive license to Springer Nature Switzerland AG 2022
P. Molander, *The Origins of Inequality*,
https://doi.org/10.1007/978-3-030-93189-6_10

institutional parameters, and the problem of hidden variables haunts any analysis of his type. Some analyses of post-war economic growth in a large number of countries have led to the conclusion that common-law countries have offered a superior legal framework that has been beneficial to economic growth primarily by limiting the public sector.[2] But, as Pistor points out, the pivotal role played by the legal profession in the development of the legal framework of capitalism has resulted in a concentration of financial, and a fortiori political, power in the British and American metropolitan areas, so superior financial and economic performance may have other roots than a possible limiting effect on government.[3]

The risks of a clinical approach to institutional policy issues become apparent when institutional solutions are transferred between countries. Institutions consist not only of codified rules of behaviour but also of the people that interpret these rules and implement them according to standard rules of behaviour.[4] This explains some of the failures in exporting legal and administrative solutions to developing and transition countries.[5]

The appropriate conclusion from these and other failures is not that the integration between economics and institutional perspectives should be discarded. Rather, it should be extended and deepened to include other parameters—trade union membership and organisation, sociological data and information from other sources. The distribution of power is particularly well suited for such analyses, with obvious relevance to the problem of inequality.

10.2   Measuring Inequality

Discussions about inequality are often marred by a lack of precision, wittingly or unwittingly. In a statement about inequality, it is necessary to specify not only what dimension of human well-being is referred to but also what measure is being used. As illustrated repeatedly in previous chapters, the picture drawn up is to a considerable extent determined by these choices.

The dominance of the Gini coefficient as a measure of inequality is unfortunate. This index is a blunt tool; societies with the same Gini coefficient for income or wealth may be very different. By way of illustration, the Gini coefficient for market incomes in Sweden was almost constant, around 0.5, between 1995 and 2018.[6] This apparent lack of change conceals large swings in the income distribution, to wit, large increases in capital incomes concentrated in the uppermost strata, balanced by increases in wage incomes in the strata just below the median. In the absence of a

---

[2] See, for example, Mahoney (2001).

[3] Pistor (2019), Chap. 7.

[4] Rothstein (2021).

[5] For a discussion of these problems, see, for example, Schick (1998).

[6] Data from Statistics Sweden.

more detailed description of the changes that occurred, one is left with the impression that nothing much happened to the income distribution. When comparisons are made between countries, even larger differences may go unnoticed.

To this may be added the fact that the Gini coefficient, like most other measures used in the literature, is relative. A 10 per cent increase in incomes across the population leaves the coefficient unchanged, but the increase in room for manoeuvre is very different if one has a monthly income of 2000 euro or 200,000 euro before the change.

The analyses of the links between inequality and growth presented in Chap. 9 show that a detailed picture of the income shares of the strata is necessary for understanding the real economic effects of inequality. Likewise, the dynamics of trust and legitimacy in society is determined mainly by what happens in the lower strata of the distribution of knowledge and incomes.

In summary, maintaining disaggregated inequality measures, such as quintile or decile shares, is a suitable and relatively easy way of increasing the precision and quality of analyses of inequality. Further, descriptions in absolute terms are a valuable complement to descriptions based on relative measures.

10.3 Modelling Approaches

The choice of models, as should be obvious from previous chapters, has important consequences for what conclusions can be drawn. This is of obvious theoretical interest—the analyst should be aware of the limitations imposed by a particular choice of model—but may also affect policy conclusions in a non-trivial way.

Rational Actors
The dominant paradigm in economic model building, general-equilibrium theory, is based on assumptions of full information and rational action. These assumptions are unrealistic, and attempts have been made to extend previous analyses to situations characterised by incomplete and asymmetric information, bounded rationality, and so on. Nonetheless, the dominant approach in models used in ministries of finance and other policymaking institutions is still based on general equilibrium.

This limitation is problematic from a policy point of view. As illustrated from a wide range of policy areas in previous chapters, individuals and households are more or less well equipped not only with financial resources but also with information, knowledge and decision-making capacities. If policies are based on the assumption of full information and rational decision-making, there is an obvious risk of bias against households that are further away from this ideal. Such bias may increase inequality even in situations where the pronounced goal of public interventions is the opposite.

Physical Analogies
Occasionally, references to physical concepts have been made in previous chapters. For instance, in a section on group interaction, equations are simplified by assuming that interaction with other group members depends not on individual relations but

only the group average—the so-called *mean-field approximation*.[7] Analogies between the natural and the social sciences are not a new phenomenon. The founders of marginalist economic theory in the second half of the eighteenth century were clearly inspired by the successful use of mathematical models in physics.[8] From the natural-science perspective, physicist Ludwig Boltzmann hypothesised that the discipline of statistical mechanics, of which he was one of the founders, might also be applicable in the analysis of social systems[9]:

> Let me just in passing indicate the broad perspective that opens up when we think about the application of this science to the statistics of living beings, human societies, sociology etc., and not only to mechanical particles.

Föllmer appears to have been one of the first to have taken up this thread, in the 1970s.[10] This approach has developed into a field of its own, *econophysics*, offering a potential for intellectual exchange across disciplinary boundaries. Applications to economics can be found in theories of, inter alia, business networks, finance, and income and wealth distribution.[11] Interaction between the disciplines has not been quite as fruitful as might be expected, however. One reason is that statistical methodology has lagged behind modelbuilding in econophysics. Authors have sometimes been content with visual inspection of diagrams in order to conclude that a power law has been identified.[12]

A second reason put forward by economists is the alleged lack of viable theoretical underpinnings of econophysical models. This criticism is somewhat hard to understand against the backdrop of the methodological tradition of economics. This discipline itself has been the target of critical comments from other social sciences because of its reliance on totally unrealistic assumptions on rational behaviour among consumers and producers. The standard line of defence has been that the assumptions underlying a theory are less important as long as it reproduces observed behaviour. The most-often cited text carrying this message is Friedman's *The Methodology of Positive Economics*, in which he developed the *as-if* argument: what really governs the behaviour of consumers and producers is immaterial as long as they behave as if they maximise utility and profit, respectively.[13] This as-if argument has its own problems. It would require the elimination of behaviour that is irrational in the economic sense of the word, but both theoretical arguments and empirical evidence show that such behaviour will prevail under mild conditions.[14]

---

[7] Section 6.3.

[8] See Sandmo (2011).

[9] Boltzmann (1905), p. 361 (my translation).

[10] Föllmer (1974).

[11] Early surveys can be found in Durlauf (1996, 1999). A comprehensive presentation of applications to income and wealth distribution is provided in Chakrabarti et al. (2013).

[12] See Newman (2005) on statistical problems associated with power laws.

[13] Friedman (1953). Similar arguments had been put forward in Alchian (1950).

[14] Hofbauer and Sandholm (2011) prove that dominated strategies may survive natural selection in a competitive environment. See also the discussion in Weibull (1994).

More importantly, consistency would require a similarly permissive attitude to be taken vis-à-vis the econophysical approach; it is useful as long as it produces valid results.

Part of the heated debate can probably be ascribed to a lack of interest in acquainting oneself with the writings of the authors of the other discipline. Objectively, there is reason to believe in the potential for fruitful cooperation, not least in the field of income and wealth distribution.[15]

10.4 Coda

David Ricardo, in his *Principles of Political Economy and Taxation*, declared that the determination of the laws of distribution was "the principal problem in Political Economy". The economic discipline that has evolved during the two centuries since the publication of Ricardo's treatise has not responded to his challenge. One of the main reasons is doubtless that many economists have considered the problem of distribution to be too value-laden, and that it should be left to the political sphere. Robert Lucas has expressed this posture with some emphasis: "Of the tendencies that are harmful to sound economics, the most seductive, and in my opinion the most poisonous, is to focus on questions of distribution."[16] Such an attitude is somewhat unusual; most economists have aspired to a more neutral position.

In recent decades, the situation has changed. As much of the research presented in this book testifies, the distribution of income and wealth is not just a matter of political tastes but has real effects on the way the social and economic systems work. Organisations such as the IMF and the OECD, which previously avoided issues of distribution and concentrated on economic growth, have acknowledged the existence of important links between inequality and growth and now spend time and energy on analyses of links between the two, as well as on the design of what is labelled *inclusive growth*.

Further, it has been shown repeatedly that the causal links between inequality and the behaviour of the economy do not confirm what was previously taken for granted before solid empirical investigations had been carried out. Inequality tends to be harmful for growth, not beneficial, and this is true at both ends of the income and wealth distribution. Social mobility is not stimulated by inequality, but hampered. And the nexus between inequality and trust provides yet another argument for why equality of both opportunities and outcomes should be a political target.

In summary, inequality represents a social phenomenon that is both theoretically challenging and politically relevant. Ricardo's dictum about the laws of distribution as the main problem in political economy remains as pertinent today as it was two centuries ago.

---

[15] See Jovanovich and Schinkus (2016) for a balanced discussion with reference to finance theory.
[16] Lucas (2004).

Bibliography

Acemoglu, D., & Robinson, J. (2006). *Economic origins of dictatorship and democracy.* Cambridge University Press.

Alchian, A. A. (1950). Uncertainty, evolution, and economic theory. *The Journal of Political Economy, 58*(3), 211–221.

Boltzmann, L. (1905). *Populäre Schriften.* Verlag von Johann Ambrosius Barth.

Chakrabarti, B. K., et al. (2013). *Econophysics of income and wealth distribution.* Cambridge University Press.

Durlauf, S. N. (1996). Statistical mechanic approaches to socioeconomic behavior. *NBER technical working paper* 203. Cambridge, MA: National Bureau of Economic Research.

Durlauf, S. N. (1999). How can statistical mechanics contribute to social science? *Proceedings of the National Academy of Sciences of the USA, 96,* 10582–10584.

Föllmer, H. (1974). Random economies with many interacting agents. *Journal of Mathematical Economics, 1,* 51–62.

Friedman, M. (1953). The methodology of positive economics. In M. Friedman (Ed.), *Essays in positive economics.* Chicago University Press.

Hofbauer, J., & Sandholm, W. H. (2011). Survival of dominated strategies under evolutionary dynamics. *Theoretical Economics, 6,* 341–377.

Jovanovich, F., & Schinkus, C. (2016). Breaking down the barriers between econophysics and financial economics. *International Review of Financial Analysis, 47,* 256–266.

Lucas, R. E. (2004). The industrial revolution: Past and future. In *2003 Annual report.* Minneapolis, MN: Federal Reserve Bank of Minneapolis.

Mahoney, P. G. (2001). The common law and economic growth: Hayek might be right. *The Journal of Legal Studies, 30*(2), 503–525.

Newman, M. E. J. (2005). Power laws, Pareto distributions and Zipf's law. *Contemporary Physics, 46*(5), 323–351.

North, D. (1990). *Institutions, institutional change and economic performance.* Cambridge University Press.

Persson, T., & Tabellini, G. (2003). *The economic effects of constitutions.* MIT Press.

Pistor, K. (2019). *The code of capital. How the law creates wealth and inequality.* Princeton University Press.

Rothstein, B. (2021). *Controlling corruption. The social contract approach.* Oxford University Press.

Sandmo, A. (2011). *Economics evolving.* Princeton University Press.

Schick, A. (1998). Why most developing countries should not try New Zealand's reforms. *The World Bank Research Observer, 13*(1), 123–131.

Weibull, J. W. (1994). The 'as if' approach to game theory: Three positive results and four obstacles. *European Economic Review, 38*(3–4), 868–881.

Printed in Great Britain
by Amazon

45113327R00126